THE MAN WHO BURIED NELSON

THE SURPRISING LIFE OF

ROBERT MYLNE

PENTONVILLE
Reservoir
NEW RIVER HEAD
OLD STREET
HOLBORN
Lincolns Inn Fields
British Museum
FLEET STREET
FLEET MARKET
WEST SMITHFIELD
BRIDGE STREET
BLACKFRYERS BRIDGE
WESTMINSTER BRIDGE
T H A M E S
BLACKMAN STREET
Philanthropic Reform

THE MAN WHO BURIED NELSON

THE SURPRISING LIFE OF

ROBERT MYLNE

ROBERT WARD

TEMPUS

For Emma, who suggested it.

Cover illustration

Blackfriars Bridge and St Paul's Cathedral by William Marlow, around 1788. (Guildhall Art Gallery, City of London)

Frontispiece

Central London, showing Blackfriars Bridge and its approaches after Mylne's embankment was complete. While building the bridge in 1760–69 Mylne lived in Arundel Street, west of the bridge between Temple Gardens and Somerset Place. He then moved to the Water House on the hillside at New River Head near the top of the map, where he lived until his death. The outline of the house, which had views across London, is shown at the south edge of the central circular reservoir just below the words 'New River Head'. It can be seen in illustration 2. (Wallis, 1802. Guildhall Library, City of London)

By the same author

London's New River (2003, reprinted 2007)

First published 2007

Tempus Publishing Limited
The Mill, Brimscombe Port,
Stroud, Gloucestershire, GL5 2QG
www.tempus-publishing.com

British Library Cataloguing in Publication Data.
A catalogue record for this book is available from the British Library.

ISBN 978 0 7524 3922 7

Typesetting and origination by Tempus Publishing Limited
Printed in Great Britain

Contents

List of Illustrations

Integrated illustrations

Picture section between pages 96 and 97

Preface

My interest in Robert Mylne developed while writing an earlier book about London, in which he played a part. Long before that was finished I began to take a closer look at what was known of his surprising life, little realising how complex the story would turn out to be or quite how unusual his personality was. Just like the strands of a rope, aspects of his behaviour pass out of sight at times but then reappear, consistent and unvarying down the years. Trying to understand what made him behave in the way that he did has been like struggling with a difficult puzzle, and one is left thinking that there may be more yet to be discovered, lurking in some box of uncatalogued letters.

A full and detailed analysis of Mylne's long career and many works is overdue, but would fill several volumes and require complex skills; in the meantime this book is an attempt to chart the course of his personal and professional life and to consider some of the more dramatic events that marked it.

For help and advice along the way my thanks are due to Mrs Audrey Mylne, Professor Bernard Rudden, John Richardson, Ian Sutton, Dr Denis Smith, Sue Hayton, Robin Winters and Peter Guillery. I also thank my colleagues and clerks, who were very tolerant of some lengthy absences from practice.

It would have been impossible to prepare this book without the unfailingly helpful staff and vast resources of libraries and archives, many of whom have allowed me to quote from original manuscripts in their collections. In particular I have been greatly assisted by Archivio Storico, Accademia Nazionale di San Luca, Rome; Birmingham City Archives; Archivist to Blair Castle, Perthshire; British Architectural Library; British Library; British Museum, Department of Prints and Drawings and Department of Coins and Medals; Corporation of London, Guildhall Library, Maps and Prints and Manuscript sections; Courtauld Institute of Art (Conway and Witt libraries); East Sussex Record Office; Edinburgh Public Library; English Heritage (Survey of London); Gloucestershire Record Office; Hertfordshire Archives; Holkham Hall; Institution of Civil Engineers; Inner Temple Library and Archives; Lambeth Palace Library; London Metropolitan

Archives; Bedfordshire and Luton Archives; Middle Temple Archives; Museum in Docklands; Museum of London; National Art Library; The National Archives (Public Record Office and Library); National Archives of Scotland; National Library of Scotland; National Maritime Museum (Caird Library); Norfolk Record Office; Orkney Library; Royal Academy; Royal College of Physicians; Royal Institute of British Architects; Royal Society; Royal Society of Arts; St Paul's Cathedral Library and Fabric Archive; Science Museum Library; Soane Museum; Library of the Society of Antiquaries of Scotland; The Birmingham Assay Office; The Sutro Library, a branch of California State Library; Thames Water; University of London, Senate House Library; University of Nottingham, Hallward Library; Victoria and Albert Museum; Warwickshire County Record Office. Illustrations show copyright credits; those unaccredited are from my own photographs.

Above all my thanks are due to Theya for her tireless enthusiasm, encouragement and critical judgment.

Robert Ward
December 2006

Simplified Pedigree of Robert Mylne FRS (1733–1811)

John Mylne *c.*1460–1513, Master Mason to James III and James IV of Scotland, the father of
Robert Mylne, Provost of Dundee, the father of
Thomas Mylne, Master Mason to Queen Mary, the father of
John Mylne d.1621, Master Mason to James VI and I, the father of
John Mylne d.1657, Master Mason to Charles I, the father of
John Mylne 1611–1667, Master Mason to Charles I and II, the uncle of
Robert Mylne 1633–1710, Master Mason to Charles II, William & Mary and Queen Anne, the father of
William Mylne 1662–1728, Mason, the father of
Thomas Mylne *c.*1705–63, Mason, who married Elizabeth Duncan, who died in 1778, and who bore him five children who survived to adulthood, namely:

Robert Mylne FRS, architect and engineer, 1733–1811, brother of
William Mylne, architect and engineer, 1734–90
Elizabeth, who died 1794 (having married Robert Selby in 1759, and whose eldest daughter, Jane, married Admiral Sir Charles Thompson Bt, who died 1799)
Jean, who died 1767
Anne *c.*1745–*c.*1820 (who married Sir John Gordon, who died 1794)

In 1770 **Robert Mylne** married Mary Home (1748–97), who bore him:
Maria 1772–94
Emilia 1773–98
Harriet 1774–1823
Caroline 1775–1844
Robert 1779–98
William Chadwell Mylne, architect and engineer, 1781–1863
Thomas, born and died 1782
Charlotte 1785–after 1811
Leonora 1788–after 1823

I

Edinburgh

Robert Mylne was born in Edinburgh on 4 January 1733, the oldest surviving son in a family of two boys and three girls, five others dying in childhood. He was named after his great grandfather who had been master mason in turn to King Charles II, William & Mary and Queen Anne, and who worked on castles, bridges and waterworks. His father Thomas was a well-established architect and stonemason, two of his paternal uncles were captains of merchant ships and an aunt was married to a schoolmaster in Leith. Other relatives were scattered from Scotland to Barbados.

Edinburgh at that time was still confined to the old town that stretched from Holyrood up to the castle. Just two miles from a good anchorage in the Firth of Forth, its spectacular site was the result of long geological processes ending in an ice age that left a gently sloping but steep-sided ramp of glacial debris a mile long, with a scoured valley either side of it and its highest end against the side of an isolated volcanic crag. It is what geologists call a crag and tail, and made the ideal site for a fortified town.[1] The castle was on the crag, approached by a single main street leading up the spine of the tail, lined with steep closes and wynds at right angles to it, where rows of old houses straggled down the sides of the hill. It was a teeming, noisy, smoky, dirty medieval city of forty thousand inhabitants, crammed into tenement buildings some as high as twelve storeys, its main street cluttered with markets. At night, when the streets were mostly empty, the residents threw their slops and waste from windows for scavengers to clear in the morning. It had a university, law courts, churches, schools and publishing houses, but there were almost no public buildings for trade, and no space to build any. It was often a drunken city, as most of its business was transacted in the hundreds of licensed taverns. There was no manufacturing, and local government was run by a medieval system of self-electing tradesmen burgesses who parcelled out public contracts among themselves but often gave good value in return.

The city had lost some of its purpose after the union of Scottish and English crowns in 1603, and this became more marked when the Scottish parliament was dissolved in 1707 and caused a migration of Government jobs to London. For this reason, there was little new building, and this affected the family fortunes. Robert's father, Thomas, was deacon of the city's stonemasons, and a portrait shows him as handsome, bewigged and exuding confidence (see illustration 7), yet there is no major building credited to his name. He is known to have built the old Infirmary, extended the Mint and designed houses for others to build as well as building them on his own account, but he was in a small way of business, and it may be that his talents were equally modest. Still he managed to support and educate his family, and when Robert was nine he was sent to the High School, arriving just as his future rival John Adam was leaving. Here the boy learnt Latin, the school's main subject, and the firm regular handwriting that makes his surviving letters a pleasure to read. His last two years were under the rector John Lees who was said to be 'severe and intolerant of dullness, but kind to more promising talents', a man who had raised the standards of the school to the point where a series of glowing reports on the annual public examinations had brought pay rises for him and his staff. Robert's brother William, who was a year younger, also went there in his turn.[2]

Earlier Mylnes had left their mark on the city. In 1481 John Mylne arrived from 'the north countrie' to become master-mason to King James III of Scotland, and six succeeding generations of the family held the same title until the reign of Queen Anne, as can be seen from the family tree on page 9. It was a John Mylne who built the Tron church, still prominent halfway along the Royal Mile, and a Robert Mylne who rebuilt the palace of Holyroodhouse for King Charles II in the 1670s, the ground floor with Doric, the first floor with Ionic and the second with Corinthian columns and entablatures, 'all of them having their true proportions, ornaments and projections of fine well cut and joined stone' as the contract required.[3]

The family lived in the centre, in Halkerston's Wynd, one of the steep narrow lanes on the north side of the Royal Mile just east of the Tron church. A stone over the entrance stair to the 'land' – a block of houses – recorded that it had been built in 1715 by the boys' grandfather William.[4] The whole wynd and its buildings, swept away in Victorian times, led down to a postern gate that gave access between the city and the end of the North Loch, so it was a minor thoroughfare. Thomas also owned the dilapidated old mansion of Powderhall near Leith, but it seems that the family usually lived in town, where the central position of Halkerston's Wynd must have been as convenient for Thomas's business as it was for the boys' schooling.

Scottish land law made it much easier than English for apartments in different parts of the same building to be sold and owned outright. This combined with the limited availability of building land on Edinburgh's constrained site meant that housing there developed in its own way. Said to have been invented by a Mylne, a pattern emerged of tall tenement blocks with grand apartments on the middle floors and poorer ones above and below, many in different ownership, with common staircases shared by all. Rich, middling and poor lived in the same buildings, and did so as members of a highly structured society where each had his place, and even the humblest were entitled to respect. One such building that survives is Mylne's Court in the Lawnmarket, built seven storeys high in 1690, and now a hall of residence.

The classic Pevsner guide to the buildings of Edinburgh lists eight Mylnes in its index. As a measure of the slow decline of their fortunes in the first half of the eighteenth century, Robert's great grandfather merits twenty mentions, his grandfather has but one, and his father Thomas none at all. Partly this may be due to demolition – only surviving buildings are listed – but it also reflects a greatly reduced output caused by the city's decline once it lost its parliament.[5]

In the course of a long life some aspects of Robert's character stand out that must have been formed in his youth, and one was a fierce pride in the reputation and achievements of his ancestors, whose name was to be seen all over the city. Down at the Palace of Holyrood an inscription on a courtyard wall still records the Robert Mylne who built it in the 1670s, as does a fantastically carved stone sundial in the garden that his father John made to celebrate the coronation of Charles I. Up at the castle, one of the fortifications was Mylne's battery, and dotted round the town were buildings like Mylne's Court and Mylne Square. Nearby in Greyfriars churchyard is possibly the finest collection of seventeenth century monuments in Scotland, some carved by Mylnes in the course of their business. Their own monument still stands 10ft high, conspicuous against the churchyard wall next to the entrance. Behind an iron paling, and flanked by Corinthian columns, cherubs and heraldry, a stone lion holds a tablet lettered in Latin and English. Below, slab upon blackened slab, the achievements of the Mylnes are set out. Most are still legible, thanks to the efforts of later Mylnes – Robert among them – who have had them recut each time smoky air has combined with wind and weather to blur the letters. There is no record of his views about this monument, but it must have been a familiar part of his childhood, and it was he who added the railing that now guards it, after the death of his parents when he became the senior member of the family.[6]

Some were remembered in verse, like John:

Great artisan, grave Senator, JOHN MILNE
Renowned for learning, prudence, parts and skill,
Who in his life Vitruvius' art had shown
Adorning others' monuments: his own
Can have no other beauty, than his name,
His memory and everlasting fame.

The Mylne after whom young Robert was named was also there:

> Here lyes the famous Robert Milne
> Laird of Balfargie, who had more skill
> In mason craft himself alone
> Than most his brethren joined in one,
> … Master Mason to several Kings of Scotland
> And Surveyor to this Citie
> Who during an active life of honest fame
> Builded among manie extensive works
> Mylne's Court, Mylne's Square and
> The Abbie of Halierud House
> Leaving by an worthy wife
> Eight sonnes and six daughters
> All placed in the world with credit to himself
> And consecrated this Monument
> To the honour of his ancestrie.

Later generations of Mylnes are remembered on the same monument, their inscriptions crowded even on to the circular columns. All this may have been the source of Robert's fondness for monumental inscriptions. Another lasting aspect of his character, an unyielding determination that things must be done correctly, may stem from the unforgiving nature of stone as a building material.

One event during his childhood must have made its impression. In 1745, when Robert was twelve, Prince Charles Edward's rebel army of highlanders took control of the city and stayed from September until November. For weeks before that the Pretender's doings had been the main topic of conversation and the main news for the press. Thirty years after his father's failed attempt to claim the crown, Charles had decided it was time to try again. It began in July when he landed from a French warship on the island of Eriskay, with a handful of companions and large quantities of arms. By August there was a £30,000 reward for his capture, as news came of highlanders flocking to join him while the British army slowly began to organise its opposition. Skirmishes and marches followed, with victories and defeats for both sides. On 7 September troops were billeted at the High School and the council gave orders for the walls to be strengthened and published a declaration of loyalty to George II, reminding him of their support for the crown during the last Jacobite rising in 1715. For many of the citizens, whatever the romantic appeal of a Stuart king, there was alarm at what the cost might be, in lives, money and disruption. Despite these precautions, the highland army burst into Edinburgh on 17 September, and although they could not hope to take the castle, which remained under English command, they held the rest of the city.[7]

A proclamation was then read to assert Charles's claims, while he installed himself at Holyrood and held glittering entertainments there – the first royal visitor since his grandfather James II sixty years earlier. On 21 September came the astonishing victory at Prestonpans – a battle so close to Edinburgh that its citizens could look out and see British troops running away to escape their victors, leaving their wounded behind on the battlefield, where they received much kinder treatment than the highlanders would meet after their defeat at Culloden the following April.

The population was divided between those jubilant at the prospect of a Stuart King and those fearful for the consequences. The Prince's charm was considerable, and it is said a third of the men and two thirds of the women were for him. Robert's mother Elizabeth Duncan was certainly one of the latter – years later Robert sent her from Paris some souvenirs of the Prince, with a letter reminding her how she had loved him. Robert's father, who as a burgess of the city had taken an oath of loyalty to King George II, was probably less keen. For him, life as an Edinburgh tradesman meant walking a constant tightrope between income and outgoings, and anything that disturbed the set routine of city life might be ruinous for many.

One of the oddest events during the occupation was the way the Royal Bank of Scotland helped Charles pay his army. The bank had removed its treasure to the castle just before the occupation, but when Charles needed to change his stock of banknotes – reluctantly disgorged by Glasgow merchants to avoid worse consequences – for gold coin to pay his troops, the bank's chief cashier went up to the castle under a flag of truce and returned with £3,000 in gold that he exchanged for the notes in a tavern in Fleshmarket Close. The money was paid over to Charles's treasurer Andrew Lumisden, and a further £3,600 was exchanged soon after, and used by Charles's army to finance its advance into England.[8] At first successful, the highlanders took Carlisle after a siege and had reached Derby when on 5 December Charles decided to turn back, after which his defeat became increasingly certain. He escaped from the rout at Culloden but it was the last time he could field an army, and by September 1746 he was on his way back to Europe leaving thousands of his supporters dead or ruined.

In Edinburgh life gradually returned to normal. When the boys' schooling finished, Thomas intended one at least to follow in his footsteps. He was growing older and needed help with the business, and his resources – in contrast to the Adam family, for example – were not such as to send his sons into the world as gentlemen architects. To this end they were both apprenticed, in Robert's case to a wright or carpenter named Daniel Wright when he was fourteen. The apprenticeship deed of 26 November 1747 provided that he could only leave Wright's service during the next six years with the consent of the Incorporation of Mary's Chapel, which regulated such matters, but it is backed by a private agreement made the same day between Wright and Thomas Mylne. This contradicts the deed by allowing that Thomas could take Robert out of Wright's service at any time, without even asking his consent, but that Wright would still discharge the indentures before the Incorporation for him at the end of the apprenticeship, just as though he had served him throughout. As a result it becomes impossible to say how much time Robert worked for a carpenter, and how much for – presumably – his own father, learning a mason's trade. His later work certainly suggests a mastery of both skills.[9] An apprenticeship may have seemed unlikely to lead to great things, yet the practical skills that he acquired in the process must have been of value later when he had to deal with the everyday problems of architecture and engineering. It may also explain the regular disputes with workmen that mark his career: perhaps he saw all too clearly when they were trying to pull the wool over their employer's eyes.

One place where Robert worked at this time was Blair Atholl, a hundred miles north of Edinburgh. Here the Duke of Atholl was having major works done to Atholl House, as post-1745 laws required him to call his castle. He had succeeded to the title after his elder brother had made the mistake of backing the Old Pretender in 1715, and also had the good fortune to inherit the Sovereignty of the Isle of Man, not yet part of the United Kingdom, which brought much revenue. His works were on a splendid scale and the interior decoration, some of it by a team of Italian craftsmen, drew freely on classical designs taken from archaeological discoveries at places such as Palmyra in the Syrian desert.

A few years later, in 1760, a visitor described the castle as surrounded with lawns, streams with ornamental bridges, stands of trees imported from America and pieces of classical statuary. A lake was set with islands that had thatched houses for swans to breed in. There was a pigeon house, a glazed summer house, and a walk lined with grotesque lead statues. Inside the castle were a beautiful dining room, a magnificent salon with a ceiling 27ft high, and a mahogany staircase, wainscoted all over with compartments for pictures, and with 'a fine frieze at each landing … of Pomeranian red deal … All the rooms … are finished in the highest manner with carvings and stucco ceilings.' Some were 'exceeding grand and adorned with costly chimney pieces of marble and exquisite carvings, some with hangings of tapestry, others with Genoa damask and marble tables …'[10]

This may have been Robert's first exposure to work of this opulent and impressive quality. He certainly did well there, as is shown by letters years later when he was in Rome. Thomas had written to tell him that the Duke wanted to know when he would be back, and Robert's reply – that he had more pressing things to do – was accompanied by a draft letter to the Duke that he asked his father to forward. It implies that he had been offered the position of head carver, so it seems he must have shown considerable skill.

One other aspect of Edinburgh life that may have influenced his future was freemasonry. Working stonemasons in Scotland had for centuries organised themselves into lodges that met regularly. Although primarily composed of masons, they sometimes invited others to join, and there was no secrecy about membership – they turned out *en masse* to join public parades on civic occasions. From this craft structure a wider freemasonry was developing, more open to outsiders. The Mylnes were long established in freemasonry, and when Robert was twenty-one Thomas introduced him as a member of his lodge. There is no evidence that freemasonry was important to Robert in later life, but it may sometimes have provided him with introductions that eased his progress.

This is all we know about the first quarter of Robert Mylne's life. At the end of that period an observer might have thought him poised to follow his father's footsteps into respectable Edinburgh obscurity.

2

Paris

Almost everything known about the Mylne brothers suggests that Robert was the leader. He was a year older, more assertive, more energetic and more resolute. So it is surprising to learn that it was William who first broke away from home and by November 1753 was in Paris (see illustration 8). Our knowledge of the years that followed comes largely from letters home, first by William and later mostly by Robert. They are far from complete: some were discarded or lost, some never arrived and when William eventually set out for home he burnt all his papers to conceal his British nationality, because the Seven Years War had begun and he feared capture. We are left with about forty letters, two thirds by Robert, spanning almost six years. Sometimes there are gaps of many months, but enough remains to build up a vivid picture.[1]

They show that at the age of nineteen William was in Paris, living on the Rue de Seine and studying drawing at the École des Arts under Professor Blondel who, he later wrote, had made him a great many fine promises which came to little. Later Robert Adam, always oddly jealous of the Mylnes, would say that William had been in France long enough to pick up some of the 'abominable French taste'.[2] William's studies were part-time, as his father expected him to supplement his meagre allowance with paid work, and he wrote that he had found a place 'in the best shop where the finest work is done'. There he had learnt the skill of cutting lettering in marble and believed he could work as well as most of the Parisian marble-cutters.

Money was in short supply, and in December 1753 he complained to his father that he was very much surprised that funds had not arrived, especially as it was near the end of the year when paid work was short because of the holidays. Meanwhile he took advantage of such free spectacles as the city offered, like a walk to the Bois de Boulogne to see Louis XV and his nobility hunting deer to the accompaniment of French horns. He was homesick, and once wrote to his father that 'it gives me a great deal of sorrow that you think so seldom of me as not to favour me with a line or two – I am here in a foreign country without friends and when letters is not answered it gives the greatest pain…' Unselfconsciously, the letter has the draft of another on its back, addressed to an unnamed 'dear love', in which he describes the perils of his brave journey to France, through 'many rough storms and tempests' and across dangerous seas.

In the spring of 1754 William spent 'two or three shillings' taking a boat down the Seine to Rouen to join James Nevay, another young Scot who was studying architecture and figure drawing there. He asked his father for funds so that he could stay as long as Nevay, saying he would be much improved by it, and that his ambition was to be something in the world, 'and not in the low way as most of the tradesmen in Edinburgh are'. He also mentioned Robert in passing, offering to get him any books he needed 'for his carving', as there was nowhere better than Paris in that regard.

In June he wrote that Nevay had advised him to study architecture with a view to setting up in practice in London, but reminded his father that as he had no friends there it would have to be on merit. Now he was finding there was more to the subject than he had ever expected. Then came what he called a bold suggestion, spurred on by his ambition. Would his father support him to study, first at the Royal Academy in Paris for a year, and then for a year in Rome? He could live in Rome for half the cost of Paris and £40 a year would be ample. The masters there were just as good and it would leave him fit to settle and practice anywhere. The expense would soon be repaid. His father should speak to others for their advice, and there was no need to tell them about the family's shortage of money, or the fact that he planned to reach Rome on foot. If all that was out of the question, how about just a trip to Rome? That could be done for £15 or £16. Meanwhile he could not take on paid work at present as it was 'so terrible hot', and even if he did he would only earn a crown a week as the French had so many holidays.

What also emerges from the letters is that Robert, the older of the two, was busy working on the Duke's improvements at Blair Atholl and was probably expected to take over his father's business in due course. With a continuing recession in the Edinburgh building trade, there would be no work for William and he must cast his eyes further afield, hence the talk of London. Never at this stage do the letters hint at any possibility that Robert might join him in France.

Unfortunately none of Thomas's letters to William have survived, but some kind of positive reaction can be surmised from William's later letters, and it seems that Robert was the one to take the next step. At the beginning of July 1754 William wrote excitedly to a wigmaker friend that he had heard from Robert and could hardly keep his seat in the coffee house when he read the letter. Robert was coming to join him – something he would have suggested long before, but never thought his parents would agree to. He even wavered on his plans for Rome, saying that he and Robert could live in Paris very cheaply, and they could always go to Rome the next year if their father's business improved.

Two weeks later he wrote to Robert and it seemed by then that Thomas was even willing to let them go to Rome, and 'as to the money my father proposes' they could live for a year on that. When he wrote to the wigmaker he had said that Robert would only need to bring a dozen strong shirts 'for their washing plays the devil with fine, and other things he will get cheaper here.' Now he added further needs – four or five dozen good black lead pencils, a couple of pen-knives, and books of French and Italian grammar as the French ones were either inferior or far dearer. Travelling, Robert should sail to Boulogne and then take the carrosse, a stage-coach, to Paris as his lack of French precluded any other method. Once in Paris he should find the Café de Conti by the Pont Neuf, and William would leave word for someone there to take him to his lodgings in the Rue de Seine.

He also set out what their next movements should be. Robert, unused to walking, could make the three-hundred-mile journey to Lyon by carrosse, while William would walk to save money. Then they could take a boat 180 miles down the Rhône to Marseille, and then a felucca to Civitavecchia, just twelve leagues from Rome. Once there they would do well for themselves, and 'by the time we come home we may put ourselves in a way above the malicious tongues of the people of Edinburgh.'

Robert's journey to Paris was not an easy one. He took ship at the port of Leith, less than two miles from home, but there were storms as they sailed down the east coast and on 4 October 1754 he wrote home from Yarmouth, where they had put in for repairs to the rigging. He had been terribly seasick, and was concerned that William would be worrying about him. It had probably been their plan to set out for Rome before winter set in, and meanwhile William in his turn wrote home about his great uneasiness at Robert's non-appearance, saying he was ready to leave the moment he arrived.

That rapid departure never happened, and the next letter, to which both brothers contributed, was from Paris on 30 November. Whereas earlier ones had all been addressed to Thomas, this was to their mother. It accompanied a parcel of gifts, silk muffetees – a kind of scarf – said to be the best in Paris for their sisters, and some mementoes of Bonnie Prince Charlie for their mother. The boys had been in touch with some of the Pretender's supporters who were still in Paris, although he had been asked to leave some time before, and there was a buckle from the blue velvet suit in which he had landed in Scotland in 1745, with a lock of his hair and a piece of his coat. As Robert wrote '… all which I am sure you'll have a great veneration for, when you love the Originall so much.' To his mother he continued with some lines of real gratitude that suggest she had pushed her husband to approve the journey. He had hardly escaped the last dangerous voyage, wrote Robert, but now they were ready for the much longer one that was the summit of their wishes. He said everyone there approved their conduct, and 'your Spirit in letting us go Surprises them all, which they justly Observe is far

Superior to the vulgar Multitude of Edin'r. I hope that our endeavours will be crowned with happiness ... to gladden your old Age and Render you happy if we can, who gave us birth.' He added a note asking that his sister Jeany should be the scribe for letters from home 'because it will not expose our circumstances so much being in her hand'. Their post was sent to a café where others might see it, and it is clear that the Mylnes were struggling to keep up appearances.

William added his own note to the letter, to say they would set out in a few days, and hoped they would be able to ease their mother's circumstances when they returned home. There was one last letter from Paris, written by Robert to his father on 4 December, mostly on money matters but also mentioning the gifts that had been sent separately, and ending with affectionate gratitude. The brothers made arrangements for some of their funds to be changed into Roman Crowns, and for a draft to be sent for them care of Abbé Grant, the head of the Scottish College in Rome whereby the Vatican looked after its Scottish interests.

A few days later they set out on what was probably their happiest adventure, a youthful journey unclouded by work or study. Luggage was sent ahead, and despite William's earlier suggestion that Robert might ride while he walked, they travelled together. As Robert wrote from Lyon, they 'footed' a good deal of it, but had also taken coaches when the chance presented itself, and then a boat from Chalon. Then follows a puzzling sentence: 'Dear Papa and Mama, I know you'll be a little uneasy that we are running thus so far off without consulting you about it, but I beg it of you to support this one effort for to appear something in the world.'

What can this mean? Had Thomas not agreed they could go to Rome? Was their mother in on the secret but not their father? The letter certainly does not square with an agreed journey. She was more indulgent of the boys than Thomas, and it is a feature of later letters that Robert first scolds his father about dilatory remittances, and then addresses himself to her for more luxurious needs, such as the price of a few decent shirts.

At Lyon the brothers were delayed waiting for their trunks to arrive, and then had a comfortable ride by passenger boat for two and a half days down the Rhône to Avignon. These boats, *diligences par eau*, seem to have been the most pleasant and sometimes the fastest way to travel where there were suitable rivers. According to an English naval officer who wrote a guide to France they were most comfortably constructed, high enough below decks to stand upright and with sash windows on both sides. Most had no sails but were drawn at a fast rate by up to ten horses, some continuing all night, while others made regular stops for meals and to sleep. The French ladies who might be met on board were extremely agreeable and entertaining, but he warned of the doctors' bills that might ensue from their friendship.[3]

At Avignon they fell in with some of the Prince's retinue again – Robert calls him the p----e in his letter – who showed them around the papal enclave where the Pretender had stayed for a while after leaving Paris. From there they travelled overland to Marseille, and had to wait there about two weeks for the next boat onwards. This gave them time to walk thirty miles to the French naval base at Toulon, and Robert made the effort to write a fuller description than he usually managed. The French would not let Britons view the arsenal at Toulon, but they did see some of the French war galleys, each rowed by three hundred slaves, 'the most miserable sight that ever I see'd in my life, being all chained by the legs and half naked'. He thought the buildings very grand, especially as to their staircases, and some would have been ranked palaces in Scotland, but they were poorly arranged and furnished inside, and in his view not suitable for such a hot climate.

As for Marseille it was a pretty town, commercially the greatest in France, with the tideless Mediterranean making it more like a loch side than a sea port to his eye. All nations were there, 'English, Scots, Irish, Spanish, Portuguese, Italians, Turks and Mahometans' and they had only been allowed into the *cabaret* where he was writing after confirming they were Christians. He said they had had many curious adventures. Sometimes they had ridden in noblemen's chaises or travelled with Knights of Malta, at others they tramped the roads with soldiers and sailors. The French, he thought, exceeded all others in their dress and civilities to strangers, and were 'extraordinary good company'. Now they had booked their onward passage, and the vessel would sail on the first fair wind for Civitavecchia.

This last letter from France is dated 3 January 1755, and conveys a sense of happiness and bubbling excitement. Robert was out in the wide world with no one to answer to, and he was enjoying himself greatly.[4]

3

Rome

Early in 1755 Robert and William Mylne arrived in Rome, having walked from Civitavecchia. The Italian peninsula was a patchwork of little states, some independent and some interlinked. Venice still had its Doge, Tuscany its Grand Duke, Piedmont, Modena and Parma their Dukes, while Naples was ruled by a son of the King of Spain. The Papal States stretched north from Rome as far as the frontiers of the Republic of Venice, skirting Tuscany but including Bologna and Ferrara.

Rome was the major centre of learning and though most of its classical remains were still buried, others including palaces, baths, amphitheatres, monuments and aqueducts were visible. Its more recent architecture included St Peter's and the Sistine chapel, and art was everywhere. It was a centre for secular as well as religious study, and foreigners flocked to enrol at its schools, including the Accademia di San Luca, named from the tradition that Luke the evangelist painted a portrait of the Virgin Mary, where painting, sculpture and architecture were taught, and its new offshoot the Accademia del Nudo for figure drawing, founded by Pope Benedict XIV the previous year. It was a friendly city, for Italian society welcomed respectable foreigners to its social functions, and since modest hospitality was the rule was able to entertain them without undue expense.

Here the Mylnes found lodgings at 3 via del Condotti, at the foot of the Spanish Steps. In the Piazza di Spagna they made their base at the Caffè dell' Inglese, which had decorations by Piranesi in the Egyptian style and took London papers, and it was there that their mail was sent. William was to remain for two and a half years, Robert for more than four. They were usually short of funds, because of Thomas's slow remittances, yet they were still able to achieve a great deal. In the Caffè they mingled with the wealthy British visitors who passed through on the Grand Tour, and in time Robert had learnt enough to give tuition in architecture.

They also had contact with the exiled Stuart court. Although Charles was elsewhere with his mistress, Clementina Walkinshaw, dissipating himself and his resources across Europe as his political hopes diminished, his father James – the

Old Pretender whose attempt to recover the Stuart crown had failed in 1715 – was alive and living in Rome. Because of his Catholic affiliations, the Pope helped to support him, and provided the Palazzo Muti where he lived and held court, and in official Roman parlance he was always referred to as the King of England. Meanwhile the British Government kept track of his doings and visitors through a network of spies organised through an agent in Venice.

It is clear that the Mylnes had some contacts with this exiled court although understandably it is not a subject dwelt on in their letters from Rome and technically might have been treasonous.[1] Thus a few years later, when James Boswell visited Voltaire on his way to Rome, he immediately denied that he might contact the Stuarts despite Voltaire's approval of them – although of course he did so when he got to Rome, and was very relieved when politics was tactfully kept out of the conversation.

The main public contact for the Stuart court was the Pretender's secretary Andrew Lumisden, the same man who, as Charles's treasurer in 1745, had changed the Glasgow merchants' reluctant tribute of banknotes for bags of gold in an Edinburgh tavern. After the defeat at Culloden, and Charles's later order that it was every man for himself, Lumisden had travelled south disguised first as a lady's servant and then as a poor schoolteacher. In London he rashly visited former comrades held in Newgate, but passed undetected and eventually made his way to Paris and then Rome. There he served as assistant secretary, and in due course secretary of state, to the Old Pretender, and he liked the young Mylnes well enough to write letters home praising them as 'honest young men, very diligent, and have good capacities for their profession', and he later called Robert 'a very worthy young man, whom I greatly esteem.'[2]

After the last happy letter from Marseille there is a tantalising nine-month gap before Robert's next surviving letter home. Presumably there were others that were lost in transit or after arrival, and as a result the record of their stay is far from complete. The first letter showing the brothers had reached Rome is one sent to them in May 1755 by Mr Burnet, a Paris-based Scottish merchant who had befriended them while they were in France. William had evidently written to him in April sending their news and asking him to buy them some pencils – apparently Roman ones were even worse than the French. It is clear from Burnet's letter that Nevay had also gone to Rome by that time, and that the three were living together.[3]

That September Robert wrote to his father, who had asked if they were doing some paid work. No, replied Robert, for although French and Italian workmen could not compare with their Scots counterparts when it came to technical skill in their crafts, 'for designing and drawing they exceed us by far.' As a result he was busy learning ways of measuring and drawing houses, how they should best be divided, and how to design the fronts of palaces and churches. His chief study at that time was how rooms should be ornamented, but he was also making figure drawings both from statues and from a live model, and said he had set out to acquire a mastery of these matters.

Robert also noted, in passing, a feature of Roman life that made a stark contrast to Edinburgh: 'one vice, too common with our country, is not known here, which is drunkenness – a man is looked upon as a beast who takes any more than for the support of nature… there is no publick Drinking places, and all the Rendezvous is to Coffee houses, which always keeps a Great Deall of politeness even among the common sort.'[4]

Other letters had more mundane requests, for even some of the basics of drawing were in short supply. Thus Robert once wrote home wanting a common case of mathematical instruments from John Bennett near Golden Square in London, with two dozen of his threepenny black lead pencils, a stick of his best Indian ink and a few ounces of French soft black chalk.[5] Such supplies would have to be put on some ship bound for Leghorn, and the service was never reliable; ordering them from Rome by way of his father in Edinburgh and on through a great uncle in London must have made the whole process maddeningly slow. Long after, he reported the goods had never arrived and nobody had heard of the ship they were supposedly consigned to.[6]

When Robert had left Scotland the works at Blair Atholl were still continuing, and it is clear that he had been expected to return there in due course to carry on, though in a better position. By the beginning of 1756 he seems to have made up his mind that he would not be going back, or at least not soon. He had received a message from the Duke sent by way of his father, asking about his welfare, wanting to know when he would be back and apparently offering him the post of chief carver at Blair. In reply he drafted a letter to the Duke, and sent it for his father to check and amend if necessary before forwarding. Whether Thomas had it copied and forwarded is unknown, but it shows how Robert's mind was working and is not without interest for that reason.

In it he regrets that he is unable to return just yet to take up the offered job as Carver, for all the benefit it might bring him both as to profit and as to reputation. He has so greatly improved in this short time, how much more might he not learn by continuing his studies? There was enough to be learnt in Rome to occupy him for several years. He named a young workman already at the castle who deserved encouragement, and set out some detailed advice he should be given as to the cutting of cornices and enrichments – high cornices should be excessively undercut for best effect. Indeed, if the Duke wanted to hold off the interior ornaments until he returned he would be grateful, and it would induce him to return all the sooner, as he had ideas for doing them in a quite different fashion from those which had been proposed. He was about to begin modelling ornaments in clay as the best method of developing a good taste, and would modify this knowledge with what he learned passing through Paris and London on his way home. The letter brims with easy self-assurance, and it gives an insight into Scottish society that so long as he was writing of matters concerning his trade a young craftsman felt able to address himself to a Duke in such terms. The letter also contains an odd line of prophecy: 'If I have made any advancement, it is not comparable at all to what I will do now when I have once begun.'[7]

In May 1756 the brothers went to Naples for six weeks and were duly impressed with the 'burning mountain' of Vesuvius, though Robert declined to describe it as it was not part of their study, merely referring his father to the description given in Addison's *Travels.* Of much greater interest, they had time to study the recently excavated remains of Pompeii and Herculaneum. The weather there was 'excessively hot', and he blessed his parents for the strong constitutions they had inherited, with the reflection that if ill-health resulted from any intemperance on their part they would deserve it, 'but thank God we can distinguish good from bad.' They were clearly enjoying life, but it sounds as though they had not thrown off the constraints of their background.[8]

By September 1756 William was already planning to set out for home, and wrote to explain his plans to his father. He seems to have had far less tolerance of hot weather than Robert. Just as he had found the June temperatures of Rouen too hot for work, now he wanted to be 'out of the hottest part of Italy before the heats' of the following spring, 'as my brother does not want to leave Italy so soon as I should wish'. He would spend four months in the north visiting the main cities and studying Palladio's villas, and then travel down the Rhine and on to London.

As for Robert, as the months went by an increasingly irritated note crept into his letters, and it seems that Thomas had failed to send the minimal funds agreed. Writing in January 1757, while William was still with him, he acknowledged his father's wishes for his future happiness, but how was he expected to do anything without the means? As for the Duke of Atholl, 'if his Grace expects I should move upon his imperiall command, you may write him that mon[e]y makes the mare to go … till I lye under these obligation to his Grace, I think I have reason to act according to my sentiments'. Abstinence and temperance were very great virtues, he pointed out, but carried no merit if they were forced upon one. Never mind the craving landlords and long-faced duns who pressed them for unpaid bills, much more real were the pinching effects of cold and hunger. If his father would try going without 'the Ordinary Sustenance of life for twenty four hours' then he might have some idea how they felt after two months of it. What would become of them on the homeward journey if they were left without means on the wide land of Germany where they had no friends ready to help the needy?

'For to tell you the truth,' he wrote, 'I won't stir from Rome till you answer my Demands … I Appeall to our past Conduct if ever they were Extravagant … while we perhaps are starving in some village on the Road, whatever Cause you may have Attributed the former Delays to, I see now that it was Intirely Your fault

... it appears we have had only £45 since the 31st of January last, which wants a few days of a year now, how far your word of honour needs to be supported by my Dear Mama, as I hinted above, you may see in this example, who promises us £60 a year ...'[9]

To put this complaint in context, it is useful to compare the Mylnes, apparently promised £30 a year each and having to manage with nearer £20, with the position of Robert Adam, also in Italy at this time, who was spending at the rate of £800 to £900 a year. Adam was five years older and had left the High School just as Robert began, but they had little in common. The Adam family was successful, prosperous, and well connected and Robert Adam grew up in a house with a large library of architectural books. From school he had gone to Edinburgh University, and while Mylne was undergoing an apprenticeship, Adam had gone straight into his father's fashionable architectural practice. In 1748 the father died, and Adam then ran the practice with his older brother John, with lucrative Government contracts to restore and strengthen the network of Highland forts in the aftermath of the '45, including the massive Fort George at Inverness. It was only after eight years' practice as an architect that he set out for Rome, and was able to do so in style. Once there he inevitably met the Mylnes and chose to be patronising rather than friendly, but he clearly had a grudging respect for them, perhaps tinged with envy at the spirit of enterprise that had taken them so far despite their poverty.

Adam wrote home about the brothers, who he called 'two sons of Deacon Mylne's.' One, he said, had studied in France and developed the abominable French taste, but the other 'came straight from Scotland, has made great progress and begins to draw extremely well, so that if he goes on he may become much better than any of those beggarly fellows who torment our native city.' He went on to foresee a risk – unless the Adam brothers in their turn spent a good deal of time studying abroad, the Mylnes might take some of the work that he felt was properly theirs – 'as he is poor, he will work much cheaper than we do.'[10]

As time went by, Adam had cause to mention the Mylnes again, and once more there is an anxious note. 'Their appearance is nothing ... they have neither money nor education to make themselves known to strangers, and of consequence few people know there are such lads in Rome, but as they apply very closely and undoubtedly will make considerable progress, one does not know what may be the consequence with the fickle, new-fangled, ignorant Scotch nobles and gentles who may prefer them to people of more taste and judgement.'[11]

The Mylnes for their part formed their own views of him, and when a letter from home passed on a rumour that Adam had been asked to rebuild Lisbon after its disastrous earthquake Robert was quick to reply. 'I am surprised at your seriousness about Mr Adam ... I assure you as an architect he makes no more figure than we do ourselves, for I see that a little study will make more than one family of architects in Scotland,' though he went on to concede that Adam certainly made a great figure at the Pretender's court, with his coach and pair of footmen. Meanwhile Adam, writing home in his turn, admitted that he was trying to make his circumstances grander than they were, and concealing the fact that he needed to work for a living, using the technique that '... a good lye well timed does well.'[12]

In the spring of 1757 William left Rome and began a slow journey north. He could not return the way they had come because the Seven Years War between Britain and France had begun the previous May. It would be more than three years till the brothers met again, and though they continued thereafter to keep in touch by letter, and occasional meetings, their lives and destinies diverged from that moment. Their parents were worried to hear they were separating, and must have written back to ask if there had been any trouble. No, William replied, there was no difference between them – 'I am tied to him by friendship which binds closer than brotherhood.'[13] We know much more about William's movements in the months that followed because Robert kept William's letters and carried them home in due course. William, unfortunately, decided to burn all evidence of his British identity before attempting the difficult crossing of the Alps, and so any confidences Robert may have given him are lost until February 1758 by which time William had reached Rotterdam and felt safe enough to retain his correspondence again.[14]

William left on a ship that sailed slowly up the coast to Leghorn, delayed by strong currents and contrary winds. From there he went inland by canal to Pisa, which he found entirely decayed and almost empty of population, despite the famous tower which he informed his stonemason father 'inclines seven times the length of a man's arm out of the perpendicular.' He found it too hot to walk to Florence, so went by chaise, and once there met acquaintances last seen in Rome. Florence, he noted, was also much decayed since losing its Grand Duke, with many of its citizens gone elsewhere 'on account of the great exactions and little trade.'[15] He stayed there a month, sketching and studying, and then walked over the Apennines to Bologna, having arranged for his trunk to be carried there by mule. There he stayed for two weeks making sketches, and then had the luck to befriend a merchant and his wife who let him travel in their chaise to Ferrara, also strangely deserted since becoming Papal territory. After ten days there he went on again, by river and then on foot with his trunk strapped to a donkey, arriving in Venice in September. That remained his base for expeditions to see some of Palladio's villas of which he made careful drawings, and he also walked to Padua and Vicenza.[16]

From Vicenza William wrote to Robert, and was naturally more forthcoming about personal matters than he was to his parents. Robert had sent him news of some disaster befalling a friend, and William replied that it had made him 'look back upon a great number of actions committed by me more foolish than wicked' and reflected that he might have met Delane's fate if the opportunity had been there. Quite what misfortune befell Solomon Delane seems to have been forgotten, but he survived as a landscape painter in Rome until 1812. William had found it expensive to take a trunk around with him, so he was reducing his luggage to what would fit in a knapsack. He had met Robert Adam again, after Mr Duff in Venice asked him to deliver a package to him. Adam, he wrote, 'behaved with vast civility and politeness. I observed him a little upon the inquisitive and answered his questions in as careful a manner as possible so he got little satisfaction … How heavy must a disappointment fall upon an ambitious spirit as his …' Adam had already set out for home in a hired coach and four, up the straight post road to Frankfurt, while William planned to head for the Alps on foot.[17] Both were heading for Edinburgh across a warring continent, but it is clear that William did not presume to beg a lift and equally that Adam did not offer one.

Robert must have written expressing worry about the Alpine route, for William wrote from Vicenza in December, saying that it was not the prospect of snow or fatigue that frightened him, but of soldiers and questioning. He also mentioned their old friend Nevay, asking Robert to destroy a drawing he had left for him and implying that the friendship had soured. Robert also fell out with Nevay a few months later, and wrote to William 'you must know we are no more friends. The occasion was his not finishing a drawing time enough, which I gave him to do for My Lord Garlies.'[18]

William's letters home give a vivid picture of his life as a penniless wandering student in those months. Sending Christmas greetings he told his parents:

> At night in the places where I put up to sleep I sit down at the fire amongst the country people and chat about this and that. When I am in town I endeavour to keep myself as honest like as my old remnants will allow, to gain a little respect, and to get an easy admittance to the palaces that I have occasion to draw … I frequently Sally out of my little appartement and pass the evening among the first nobility at the Expence of a Dish of Coffee, as they mostly know me by seeing me measuring here and there, and are people of taste, enquiring about the things I have seen in my travels …

Early in January he wrote again from Venice, and must have alarmed them with a description of himself as being so 'dryed by the heats of the sun and of late withered by colds that I am sure you won't know me. My back teeth are almost gone, my right eye which was always very tender is turned very dim…' Writing to Robert the same day he painted a rosier picture, saying that he had gone back to Venice over the New Year and 'took the diversion of a few plays to scour my brain, as I was almost turned misanthrope by being so much alone.'[19]

By mid-February William had reached the safety of Rotterdam, and wrote to tell Robert of his journey. From Vicenza he had gone to Montebello, then to Verona, and Peri. At Burgetto he had met with the rearguard of six thousand Germans marching for Silesia – 'you guess my situation upon finding myself alone among them cattle – as I had burned all my letters and everything that might appear English I passed as an Italian.' He had managed to get past the guard placed at the entrance of the Tyrol without examination, as they took him for a soldier or someone belonging to the troops. Then he had met up with some hussars and feared being pressed, but was lucky to get lift on a post chaise to Rovereto. There he made a bargain with a postmaster and set out on horseback for Trent through the snow. From Bolsano he went on to Innsbruck by sledge, and again by sledge to Augsburg, in constant fear of being robbed and murdered by the deserters who plagued the region, where many such were dangling on roadside gibbets. A post wagon took him on to Frankfurt where the river was frozen so he had to continue by road to Cologne. From there he reached Utrecht and finally Rotterdam, by which time he said he was 'reduced to a skeleton by fatigue and my purse to a crisis by expence.' At Rotterdam he had been lucky enough to find Mr Burnet, who had moved there from Paris when the war began, and he had helped him out. He ended by saying that he would take a Dutch ship for Scotland at the end of the week, 'and you'll write me at Halkerston's Wynd'.[20]

Once William was back in Edinburgh, Robert wrote with news of at least some of his doings in Rome. Richard Phelps, a tutor, had been a generous support to him. He had had a succession of architectural pupils thanks to introductions from Abbé Grant, most of them young men from landed families like Lord Garlies and Mr Fermor. The latter took lessons every evening and was determined to have a new house built as soon as he got home. This was, some years later, Tusmore in Oxfordshire, one of Mylne's grandest commissions. Robert was also designing a house for a merchant. He wanted to hear more details of William's journey: his last letter had too many 'wide lines, large letters and broad margins' for his liking.[21]

William duly obliged in June, having by then taken over as the Edinburgh family correspondent. He gave a much fuller description of the homeward journey, which had taken its toll on him. It had been '... one continued ague of fears and fatigues that reduced me so low that I am hardly as yet come to myself. I passed all the way as an Italian, frightened to be taken up as a spy, and dragged over mountains of snow upon sledges almost to death, always covered with hoar frost and snow.' At Verona he could hardly get a place to stay because of his rough appearance. He described his new life in Edinburgh a little, saying that business was 'very dead at present but if anything cast up I have a tolerable chance ... I have entered in Mary's chapell and set up as a mason and architect.' The chapel was a guild for Edinburgh trades, and William seems to have taken over from his ageing father. Edinburgh life went on much as usual, he said, and Robert's friends sometimes raised a glass of punch to toast his health.[22] William's letter had also encouraged Robert to come home soon, and at the end of June Robert replied:

> You press me to come home, I answer what to do, to be joviall for a month, enter a freeman wright, and sit down patiently to wait the first blessing, which God knows when will arrive, and poor I in the meantime sink into Oblivion; These are wide and enlarged prospects capable to hook any youngster ... If I judge by you I have very little prospect of anything: especially if I was to put it in execution – you as being the first settled would catch at everything. But, dear Will, there is a thing which works in my own breast more ... it is the reproaching myself with inactivity to the age of 25, and resting a burthen to the best of parents for 7 too much.

Then came a declaration and a prophecy: 'This has made me resolve, and it shall be done, the great era of my life is coming, and I shall shoot athwart your northern hemisphere very soon...' How this was to be accomplished he did not deign to say, but the letter sounds very confident, and has echoes of his similar prophecy to the Duke of Atholl two years earlier.[23]

Once he was left behind in Rome, Robert seems to have reappraised his situation and decided that his only hope for a worthwhile future was to apply himself to excellence by study. One of the contacts who may have provided inspiration was Piranesi. This artist, thirteen years his senior, had a flourishing studio where he gave tuition to Robert among others. He too was the son of a stonemason and it is clear they became friends and corresponded as such after Robert's return to England. As well as giving friendship he taught skills and almost certainly implanted his own ideas, so that his influence can be seen at times in Mylne's later designs. Piranesi had a great admiration for the magnificent achievements of classical Rome, far less for those of Greece. Robert was to write many years later that he had made a study of the Roman system of aqueducts for water supply, and Piranesi probably guided him in this. Piranesi also took a great interest in bridges, and this too may have given Robert ideas and insights that were to affect his whole life.

All the time he was in Rome, Robert paid close attention to the structure of St Peter's, and wrote later that for four years its cupola had been the focus of his study of the science of sound building.[24] Around this time he also began to design houses, the first for a Dresden merchant who was passing through Rome, introduced to him by one of the Dance brothers, but its plans have not survived and it is not known if it was ever built.[25] Other designs were sketched out for English visitors, and some eventually led to useful commissions.

Another change in Robert's life was that he increasingly taught others, as he had mentioned to William. Wealthy young Britons passing through Rome were encouraged to 'study a little architecture at Rome ... an elegant and useful acquirement, and soon learnt as far as will be useful.'[26] Robert was now competent to teach architecture, the visitors could afford to pay, and he was thus able to solve the problem of his father's slow remittances. Long periods in such company must have added to his social poise – potentially a great help in the transition from craftsman to architect. He was also prepared to travel around Italy for similar purposes. One such trip took him as far as Sicily in the company of Richard Phelps in February 1757, by way of the Kingdom of Naples, where the British consul issued a passport for onward travel. The war was under way and the consul addressed it to 'All Admirals, Commanders and other officers in the service of His Britannic Majesty, and to all Admirals, Commanders and Governors belonging to Princes and States, Friends and Allies to the Crown of Great Britain'. This seems to have sufficed for the short hop to Sicily, where they studied the Greek temples being excavated at Agrigento and elsewhere, and Robert made measured drawings later said to be of unequalled quality. He wrote to tell William that he was trying to complete his drawings of Sicilian antiquities, and the German archaeologist Winckelmann acknowledged his help in his work on Sicilian antiquities a few years later. Robert also made notes for updating the map of Sicily, and published it many years later in London.[27]

A letter written when Robert had occasion to complain at his father's delays sheds some light on that trip. 'By your delaying always since we came to Rome,' he wrote, 'you have gained eight months on us ... our necessities could never have allowed it, had it not been for my excursion into Sicily with Mr Phelps, by which I both saved and gained enough to set us free of the world.' Not totally free of the world it seems, as the letter ends with a plea to his mother for the price of half a dozen shirts.[28]

After writing his prophetic letter to William at the end of June, Robert's next letter was not until 23 September, and its structure is deliberately teasing. The first page and a half is a careful blend of trivial news, gossip about acquaintances, and his reaction to some bad news William had sent him – there had been a fire in their buildings at Halkerston's Wynd one afternoon in August, with much destruction and some loss of life, though none of the Mylnes had been injured and the buildings were insured. Robert supposed his mother would be upset about it even though she had not been there. He was surprised the fire had not continued up the Wynd, where there were old timber houses built as if waiting for a bonfire. That subject dealt with, he went on to explain how he had made a useful acquaintance in the Earl of Cardigan's son, Lord Brudenell. He reassured William that his curiosity in a previous letter about the progress of the Duke of Atholl's building work had been just that – he would never return there as Carver, though he might return as Architect. There is not the faintest hint that the rest of the letter will have any significance, and one can imagine the reader's head beginning to nod a little as he read on, perhaps even putting it aside to finish later as the candle flickered. Then comes the news like a thunderclap, and one pictures an involuntary shout in Edinburgh, perhaps even a dropped cup. Robert had just won the first prize for architecture at the Academy of St Luke.[29]

4

'... thump, thump, thump, I feel it yet...'

Early in 1758 one goal had begun to dominate Mylne's life, though his letters never mentioned it. At that period the Academy of St Luke held a prize competition every three years, with graded awards in separate classes for painting, sculpture and architecture. Plainly Mylne would be at a disadvantage against Italian students, but any kind of award could only help his reputation. The architectural subject that year was for a design for a public gallery to house the busts of distinguished men. It was a composition that lent itself to a magnificent treatment, and Robert had been giving close attention to the subject of rich ornamentation as shown by his letter to the Duke of Atholl. The perfectionist in him set to work.

Robert's letter to William announcing the result shows him understandably intoxicated with happiness and pride in his achievement. It had been a long process and then the judging had taken place in the presence of the Cardinal Chamberlain – 'when the first prize was unanimously alloted me – think on my breast when I received the news – thump, thump, thump, I feel it yet.'[1] To prepare the drawings had cost him seven months' hard work, he said, but submitting finished drawings was only part of the process (see illustrations 11–13). All candidates also had to complete a drawing in the presence of the judges, and the subject for that was not known in advance. There was then a delay because the Pope had died in May and everything was postponed. Eventually the drawings were all submitted on 6 September, and the next day the candidates attended for the *prova*, a test drawing to be completed in two hours and under public scrutiny. The subject here was a design for an elaborate altar adorned with composite columns – apparently not good news for Robert. 'I am sure you are quaking with me now, with that cursed subject,' he wrote, but his had been fine by comparison with the others. They were all judged a few days later, and he was then announced the winner.

He went on to describe the prize-giving ceremony that had followed. It had taken place in the great hall of the Capitol, specially adorned for the occasion, and he continued:

> … we received our Meddalls from the hands of 16 Cardinalls who were present. It fell to my share to have mine from the hands of Cardinall Sacropanti, who paid me a compliment on the occasion. They are 2 in number, all silver, about 5 crowns value each, with the portrait of the present pope on them, and the reverse St Luke painting the Madonna. A Monsignor made an oration on the Occasion and the Acadian Poets rehearsed Sonettos in our praises. There were likewise three symphonys of Musick, to make out the Ceremony. Beside the Cardinalls there were present the Embassadors, Nobility, Ladyes, Monsignors, and as many Artists and Deletanti as the room could hold. We *premiati* were set in the most conspicuous place, and as I happened to be the best drest, I was oblidged to bear the eyes of the whole multitude. We were called out aloud by our name and country, and then consider the situation of my heart when descending from my throne to receive my reward. When all was over I received the compliments of most of the Cardinalls & principall people, my countrymen & foreigners, which were the more agreeable as I find since no one has offered to criticise my Drawings, but on the contrary has raised them to the skies. Pardon the pride of this high tho' true description of my honours, for I wrote for my father & mother, not to you; as all my joy is to reflect honour on them & on my country, for I am the first Brittainer that has ever had a first prize here.

Reading Mylne's description of the ceremony (see illustration 14), some might be tempted to wonder if he embellished the grandeur a little. Did such massed ranks of cardinals really turn out for a student prize giving? Was there really music specially composed for the occasion? The answer is to be found in two published works – one a book written in Italian and published in Rome that year, the other a guide to Italy written later by a French visitor who happened to attend the ceremony.[2] As well as listing the prize winners, they both describe the ceremony, the hall hung all over with crimson damask and velvet edged with gold lace and fringes, and lit by candelabras and lustres, twenty cardinals rather than Mylne's sixteen, orations, symphonies and sonnets specially composed for the occasion. Everything is much as Mylne described, with one exception – the presence of the 'King of England' – the Old Pretender – in a gallery specially provided for him. His presence is something Mylne did not mention when describing the ceremony to his brother. Perhaps he felt that it was best left unsaid, as it would hardly be well received in England, where he already knew the prize could attract work. This is shown in the same letter, where he continued with the suggestion that William might arrange for a short notice in the newspapers. The press duly reported what great pleasure the news had brought, being 'so justly conferred … an instance of the uncommon merit for so young a man, to gain a prize which the greatest architects in this country with keenness aspire after.'[3] It is clear from the letter that Mylne received two silver medals as his prize. Presumably this was because he had submitted the best *prova* as well as the best design. Although the *prova* served a number of purposes, not least by making it difficult for a fraudulent entrant to submit someone else's work for the main competition, it did not follow that the overall winner would submit the best *prova*, so two medals would be justified. In a later letter he explained the reason he believed his designs had been preferred; he had '... interpreted the meaning, composed the Architecture, & drawn it with more knowledge of Light & shadow than any.' He went on to say that the prize had gained him the powerful patronage of Prince Altieri, who was already applying to the Pope for the necessary dispensation so that Mylne, a protestant, could be proposed for the further honour of election as a professor of the Academy.

Robert was alive to the intellectual benefits of the competition as well as the practical ones. As he explained to William, the nature of the competition had forced him to follow 'a strict proposition ... all difficulties must be surmounted, so that the judgement becomes more strengthened,' and he compared this favourably with undirected private study where one could simply change direction when difficulties arose.[4]

What he did not mention in the letter is that he had prepared not one design but two. Surviving copies in London and at the Academy of St Luke show that having completed all the drawings needed for one submission he went on to make a second set to a different design. Both have a monumental public building embellished with statuary set beyond a large colonnaded piazza, but they differ greatly in their details. The one not submitted – presumably the first – has what has been described as a mixture of fairly conventional Renaissance and Baroque features with some Neo-Classical elements, whereas the prize-winning entry has many more Neo-Classical features and may have been preferred, by Mylne and the judges, for that reason. Its drawings, some now in London but most still in Rome, show a colonnaded piazza topped with lines of statues giving access to an inner block lined with giant Corinthian columns, its windows showing Neo-Classical details. The interior mixes vaults, domes and varying room heights with much statuary and elaborate ornamentation (see illustrations 11–13).[5] Why were there apparently two completed designs? If he changed his mind so radically before the first set of drawings was complete, why bother to finish it? Perhaps the unexpected postponement caused by the Pope's death gave him the time he needed to start again at the beginning and resolve ideas that had been occurring to him as he worked through the first design.

In his letter to William, he set out his future plans. He would settle in London and practice there, because 'a Prophet is better believed out of his own country than in it.' Also, Edinburgh people knew he had spent seven years in just one small branch of his new profession. Further, he had more friends in England, and it was easier to get money there. His plan was to be back in Edinburgh soon, so as to arrange the necessary ways and means for achieving all this. In the meantime he would stay two months in Rome and then set out for Florence, Bologna and then Venice.

From there he hoped to get a passport that would allow him to return through France even if the war still continued, and then he would come home by way of Holland. He asked William to warn their father of the inevitable costs. He would need another year's allowance, and asked for immediate credit of £40 to await him at Leghorn, a free port established by the Medicis with long-established settlements of English, Italian, Jewish, Greek and French merchants through which much of the commerce of the region was transacted. If he was made an Academician, that alone would cost fifty-five crowns. He might not need to use all these funds, as he expected to profit from tuition for English visitors over the winter.

Although he did not mention it, Robert had also written a letter to the Academy at Parma, which held a competition with the prize of a sizeable gold medal. Mylne wrote to their secretary asking to clarify some details of what was required on 12 September, immediately after winning the prize at St Luke's and before it had been presented to him, with a view to submitting an entry that autumn. He does not mention this in his letters home, and the Parma Academy has no later details of any entry he may have made. Perhaps he decided he had enough to be going on with, or perhaps he realised that a failure there, or even a lesser prize than the highest, could only tarnish the success he had already won. If so, that can be seen as a sensible decision to leave that particular game while he was still ahead, and suggests he had matured during his years in Rome.[6]

By January Robert was able to tell William that he had been unanimously elected a professor – an *Accademico di Merito* – of the Academy of St Luke. This was not something that automatically followed from his prize, and it entitled him to vote in the Academy's deliberations and take part in judging future competitions. He was particularly pleased that he did not have to feel under any obligation to old friends for this – the Prince Altieri, who had been impressed by his prize-winning design, had arranged the necessary dispensation. There would be a ceremony in February, requiring him to obtain a black toga and black silk cloak, with white bands and his hair curling free down his back. He knew William would find that a ridiculous sight, but asked him to congratulate their parents for him and pass on his pride at having the chance to 'reflect some honour upon those that gave me birth.' Even in a moment of such pleasure he did not forget practicalities, and went on to suggest 'were you to put it in a very modest paragraph in the news, I Believe it would be apropos enough, so soon after the last. But do as you think best.' The paragraph duly appeared.[7]

He was able to pass on news of some future earnings. Richard Phelps had agreed that as soon as Robert reached London he should set to work preparing the *Antiquities of Sicily* for publication. This would be a lengthy procedure, and Robert thought it would bring him £80 a year as long as it lasted, which would be a useful basic income. He was also cultivating new contacts such as Murray and Lord Garlies, with a view to providing William with introductions 'into the great world in Scotland,' so that he too could develop his practice.

William wrote back at the end of February to say there was £30 now available to Robert. He was sorry for the delay but the misfortune of the fire had drained their funds. He had news as well. Everyone was delighted with Robert's new honour. According to Edinburgh rumour, the Pope had knighted him and an English nobleman had settled money on him. Even the London papers had reported his success with paragraphs and verses. As for William himself, things seemed to be looking up. 'I have the town's work for this year,' he reported, as well as a chance to build a bridge over the Tweed. He also mentioned a matter that must at the time have seemed a cause for rejoicing – he had been asked to make a plan for 'the New Intended North passage ... It will be a great undertaking when it is brought to bear'. This was the proposal for a bridge from the High Street over the deep valley of the North Loch to the ground where the New Town was to be built, potentially a very prestigious piece of work. His only problem was a lack of capital, and he joked that he might have to marry for money. He mentioned a Miss Hamilton who had sent her compliments, and called her Robert's sweetheart, though we never hear of her again.[8]

Robert then stayed in Rome for a few more months. His prize-winning drawings were on display at the Academy, and were still there in March 1761 when they received their only known sour review – not surprisingly from a member of the Adam family. This was James, the youngest, who went to see them with Abbé Grant. He had no great expectations of them, he wrote to his brother Robert, but people of taste had puffed Mylne as a great man, so he went to see for himself and concluded that 'if the English have Eyes, you have nothing to fear from that quarter.'[9] Abbé Grant does not seem to have shared that view, as he wrote to the wealthy connoisseur Lord Charlemont that Mylne was a young man of great worth and merit with uncommon talents and singular application, who had 'done himself and his country great honour here by having by unanimous consent and universal approbation' won the prize. In a letter that gives an insight into the way influence worked at that time Grant commended him to Charlemont's patronage, and said he would regard anything Charlemont could do for him as though it were done for himself.[10] By April Robert was in Florence, beginning a leisurely journey home in much more comfortable circumstances than William had endured, and he wrote a long and relaxed letter. He had left their Roman acquaintances merry and well. He had had a good winter there, collecting enough fees from the English gentry to leave 'loaded with Riches and honour.' The riches were not great, but he thought the honour could hardly be surpassed in his profession. He had given Sir Wyndham Knatchbull the plan of a house, and hoped to be able to build it for him in due course. He was travelling with the son of an Edinburgh advocate, who had wanted a companion for the journey and was happy to pay the travelling costs, as well as providing good discourse. They had taken a good road by way of Terni, Narni, Fuligno, Perugia, Cortona, and Arezzo. Having bought some good clothes in Rome he would be able to mix with the best company along the way. He expected to be elected to the Academy of Design in Florence. He was very glad to hear of William's growing success, and once back he could provide him with some useful introductions for he either knew, or had letters to, a good number of contacts. Meanwhile William could write to him at the French School in Venice.[11] As predicted, at Florence he was elected a member of the Academy of Design, and the same happened with the Academy at Bologna. By June he had visited Venice, where he said the English Resident as well as the Consul and their ladies had been particularly civil to him, thanks to Lord Garlies. He next went on to Brescia, and by this stage his plans to travel back with a Mr Deering had fallen through as Deering was delayed for another year. Fortunately he then met a Mr Devisme who was also heading for England and was prepared to pay the major part of the travelling costs. To ease their journey through Germany they got themselves French, Imperial and English passports, and so he hoped to avoid any delays. He had visited some of Palladio's villas, but whereas William had drawn them with great care, Robert concentrated on their situation and effect.[12] From Brescia, Robert's onward journey took him by much the same route as William's, though more comfortably. By 24 June he was writing to William from Basle, describing a journey over the Alps whose only problem had been a brief scare that he might have lost his sight after over-exposure to the sun stripped the skin from his face and made his eyes swell.[13] From Basle they took a coach for Mainz, and then a boat down the Rhine to Rotterdam, relieved to have avoided contact with the fighting forces who, he wrote, were in the country around Hesse and Paderborn. He also mentioned his plans for the last part of his travels. He would go up to Amsterdam before returning home, as he thought it worthwhile to see what he called 'a country so new and so much connected with the mechanical part of our trade', presumably the Dutch mastery of machines for pumping and channelling water that had reclaimed so much land from the sea.

At that point chance intervened, as Robert was still in company with Devisme, who decided to head straight to London. For Robert, it was almost five years since he had set out on his travels, and he probably found the proximity of England irresistible. Amsterdam was put aside and they went to the Hook of Holland where there happened to be a packet boat waiting to sail for Harwich. If chance had not intervened in that way he would probably have missed the greatest opportunity he would ever have.[14]

5

London

Crossing the North Sea took just twenty-eight hours, and on the evening of Sunday 15 July 1759 Robert stepped ashore at Harwich and found an inn for the night. He was twenty-six years old and it was almost five years since he set out from Edinburgh. The next morning he took the London coach, which reached Islington by evening, so that he eventually arrived in London on the morning of Tuesday 17 July. His plan, as he had explained to William, was to start preparing *The Antiquities of Sicily* for publication, with funding from Richard Phelps that would support him as he set up an architectural practice. He meant his future life to be carefully planned and had no intention of leaving things to chance, believing, as he had once written to William, 'the fortune of the world which we so much hunt after, is more owing to Judgement than good luck'.[1]

How mistaken he was in this view was about to become clear, with a succession of changes of plan forced on him by chance events. He had expected to make an impressive entry into London's fashionable world, buoyed up by the fame of his prize. He very soon discovered that because it was July, everyone who mattered had gone – 'all the Company are out of town at present & those particularly to whom I have recommendatory letters,' as he wrote to William the same day. At first he thought he could at least make progress with the book, but even Phelps's contact in the book trade, George Pitts, was out of town exercising with his county militia and not expected back for months.[2] Mylne's disappointment must have been enormous. He had been so keen to make his London début that he had even missed the chance to see Amsterdam. Now he knew he could have gone there and learnt new skills, but he had wasted the opportunity and, as it happened, would never cross the seas again.

A new plan suggested itself and he proposed it to William. He would 'go down to you in Scotland for some months & then come back here, if I am to settle in London, as if new come to England.'[3] So, having fluffed his grand entrance he proposed to slip away unnoticed in the hope that a fresh triumphal entry could be staged in the autumn. It would also mean that he could visit Galloway where there was a chance of designing a new house for one of Lord Garlies's relatives, as well as going home to see the family and arrange his future finances. While he awaited William's reply he could lodge with their Doby cousins in Lichfield Street.

This second – or third – plan reached an advanced stage before it too foundered. As he wrote to William in August, after receiving what must have been an encouraging reply from William to his earlier proposal, he had actually been 'aboard of a ship with an intention to pay you a visit' when he learnt something that meant it must be deferred. The coasting vessels that plied the east coast between Leith and the Pool of London were the easiest way to travel to Edinburgh and Robert must have been on one when he heard the news. In defiance of his careful planning it almost certainly came not from one of the contacts he had cultivated but from a fellow passenger or a member of the crew. It was no more than a piece of topical gossip, suitable to be mentioned to the stranger next to you to fill a silence, when a group of travellers lounge against the ship's rail and gaze up the Thames as they listen for the order to slip moorings that will begin the voyage.

The news, whoever it came from, must have electrified Robert. There was to be a new bridge right there, and a competition had just begun to choose the best design. Pure chance had landed the fledgling architect, freshly minted prize medal in his pocket, at exactly the time and place of that rare event, an open architectural competition for a major project.

Can he have delayed for more than an instant? Did he panic to get ashore? Did a crewman grumble at turning out the hold to find his bags? His letters do not explain, but it is tempting to imagine the same 'thump, thump, thump' in his breast that he had described at Rome.

We do not know the exact date on which he heard the news. It was clearly after he had received, and had time to react to, William's reply to his letter of 17 July. Allowing at least ten days for letters to go from London to Edinburgh and back again, it must have been late July or early August. By about 18 August, when he next wrote to William, he wrote not only the news that there was to be a new bridge, but also of the extraordinary progress he had made. The bridge was to be at Blackfriars and the City of London had already raised the money to build it, yet had not settled on a design. He had not only contacted the Committee for building it, but also had a supporter in its secretary, 'a man of great weight among them and very much my friend'. There was one other principal contender, who was currently out of London, and he had been written to 'by his friends and mine to see if he will agree that we should join together in it; and if so there is great probability that nobody else can have it.'[4]

Before examining how this had come about it may be useful to consider the whole position of London and its bridges. London was almost certainly chosen as the site of a Roman city because it was the lowest convenient point to bridge the Thames. The result was a city with the ingredients for success: a good harbour safely inland for inward and outward shipping, and a bridge, probably wooden, to form the nucleus of roads radiating in all directions that would make the place a focus of trade and a convenient stop for travellers. What grew there was the City of London on the north bank, administratively quite separate from the much smaller city that developed upstream at Westminster long afterwards and the small settlement opposite at Southwark. The London Bridge of Mylne's time was six centuries old, inconveniently narrow because of the shops and houses that lined both sides – two carts could not pass one another – and a constant source of delay to traffic. It was also a menace to boats on the river, as its nineteen arches blocked half the waterway at high tide. Worse still, each of the piers that supported the arches was protected by a stone 'starling' below high water mark, intended to protect the piers from collision damage. These caused an even greater constriction, because when the tide was low more than three quarters of the width of the river was blocked. This meant that water ponded up behind the bridge, and the difference in level could be almost 5ft, so that a waterfall poured through each arch, making navigation both difficult and dangerous. The risk to boats was worst at high water, when the starlings were hidden like submerged rocks. Yet until 1729, when Fulham Bridge was built, the nearest bridge to London was many miles upstream at Kingston.

The reason for this odd state of affairs was that the City, far from wanting more bridges, had a long history of opposing them. This was because of the financial benefits of controlling the river's main crossing point. Though London Bridge was narrow and congested most of its users had no real alternative, and if their passage was delayed they would probably spend more time and money in the City. In the 1660s there had been proposals for a bridge at Westminster, but the City sent a delegation to wait upon Charles II to implore him to oppose it. He did so, possibly influenced by the City's readiness to help him in return. Thus the records of Common Council for October 1664 show that the citizens lent the King £100,000, and took the opportunity to express their gratitude for 'his Majesty's good and favour towards them expressed in preventing the new bridge proposed to be built over the river of Thames betwixt Lambeth and Westminster, which ... would have been of dangerous consequence to the state of this city.' A few years later in 1671 when Putney Bridge was proposed, the House of Commons was told that the Lord Mayor believed that if it was built wide enough for carts 'London would be destroyed'.[5]

All this changed once Westminster Bridge was built in 1738–50, funded first from lotteries and then directly by Parliament. Why should through traffic queue in crowded streets leading to a congested London Bridge if it had any choice? Westminster's gain was London's loss. The City grew anxious, but bridges are expensive and there were few spare funds. In any case the whole waterfront was in the hands of many different private owners, and a bridge was no use without approach roads. This was not a simple problem – all efforts to remodel the City with wide new streets after the Great Fire ninety years earlier had foundered on the difficulty that every piece of scorched earth had an owner, who was often determined to rebuild on that spot and no other.

The eventual answer was found at Blackfriars, at the western edge of the City but still within its boundaries. It was not the smartest of areas. The Fleet River, long a receptacle for all kinds of filth, reached the Thames there and emptied itself on to large mudflats that were exposed at low tide. The Fleet had been widened and dredged after the Great Fire so that boats could be taken up to what was called Bridewell Dock, and on a high tide could load and unload as far up as Holborn, though the part north of Fleet Street was covered over in 1733. On either side of the Fleet were the haunts of the criminal poor, low taverns, jails and all the detritus of city margins, so that any land purchases would be cheap.

The City took some advice from the engineer John Smeaton (see illustration 15) on how cheaply a bridge might be built, and obtained an Act of Parliament in 1756 that gave it limited powers to borrow money for the purpose. One problem was that the Act only allowed the borrowings to be secured against future income from charging a toll on the bridge – an inherently uncertain security, which at best depended on guesswork, and which would be worthless so long as the bridge was for any reason incomplete. Then, the project being such a major one, a vast committee of thirty-six aldermen and councillors was formed to take things forward. There were long deliberations about the structure. Should it be stone or wood, or perhaps a wooden superstructure on stone piers? How could the money be raised, as the powers given by Parliament only allowed for four per cent interest to be paid and the outbreak of war had forced market rates above that? Eventually a scheme was devised to cope with these difficulties by offering subscribers the City's seal as security and agreeing to accept the money in quarterly instalments spread over six years, while interest would be paid from the time of the first instalment. The matter was then ready to go ahead and subscriptions were invited. The offer was a success, and became fully subscribed on 15 July 1759 – the very day that Robert Mylne landed at Harwich.

A public announcement was then made inviting interested parties to submit plans with estimates, with a closing date of 4 October. For those who had been following the whole proceedings for years past this would cause no difficulty, and no doubt many designs were already prepared. For Mylne, who knew nothing of it until the end of July, there were barely two months in which to produce completed designs for a major work of civil engineering across what was to him an unknown river in an unknown city. He had never built even the smallest of bridges. The only completed structure he had to his credit is referred to in a letter he wrote to his brother from Rome when he said that a house in Dresden for which he was drawing plans would be his first, 'except indeed that where you live in at present, for it will always be like my darling child, as it was the production of my earlier years.'[6] So it seems that he had designed some kind of alteration or addition to his grandfather's block at Halkerston's Wynd, but to claim that as a qualification for throwing a bridge over the Thames would have been to invite ridicule. He was almost penniless, and depended on the kindness of a relative for a place to sleep in a modest part of town. He can only have returned with very limited drawing equipment. As a Scot he was counted a foreigner in London, and Scots were widely resented. He was still in his twenties and apprenticeship to a carpenter did not make him a bridge-builder. His only conceivable qualification was that he had won a student's prize for designing an imaginary building in a foreign country, and a Catholic one at that. His competitors would include experienced architects and engineers. Chief among them was John Smeaton, a household name for works of this kind, who as everybody knew was currently putting the finishing touches to the Eddystone lighthouse he had somehow built on a tiny rock fourteen miles off Plymouth, who had just won the Royal Society's most prestigious award for experimental work, and whose design for a bridge at Blackfriars had been the basis of the 1756 Act of Parliament.

All things considered, wiser heads might have urged Mylne to get back on the boat and pay his parents an overdue visit. Instead he followed his own leanings and set to work. So far he had benefited from the accident of timing that had brought him to London that month. There was to be one more chance fact that was probably decisive.

The secretary of the bridge committee was a Scot, and he it was who befriended the newcomer. John Paterson, one of the City's elected common councillors, was a solicitor who had just become the city's auditor (see illustration 9). His grandfather had been the last archbishop of Glasgow, but the family had suffered for their Stuart sympathies at the time of the 1715 rebellion. He was thirty years older than Mylne, and was also a freemason, or so one might guess from a single diary note Mylne made years later that he had 'dined with Paterson at lodge.' If this assumption is correct it might explain how they came to speak in the first place, but it does not begin to explain why Paterson became his supporter. The design of a publicly funded bridge over a broad tidal river in one's own city is not some kind of sinecure to be entrusted to the first stranger in need of a handout, whatever his affiliations, and certainly not by an ambitious politician such as Paterson was – he was later a member of Parliament for several years.

Following the history of the bridge in the years that followed, the likely explanation is that Paterson realised he had found someone who had a rare combination of qualities: not only the basic ability to design a bridge, as well as the energy, practical skills and determination necessary to see it through to completion, but one who shared his own uncompromising beliefs as to the integrity that a public project of this nature demanded. That anyway may serve as a working hypothesis, to consider as the story unfolds.

August and September must have been very busy months for Mylne. Any bridge has to be designed for the particular river it crosses, and a full knowledge of its characteristics is needed, not just its width and depth, but matters such as the composition and profile of its bed and banks, its tidal variations and seasonal differences in level, as well as the dimensions of the river craft and road traffic that will use it. Blackfriars was not an ideal location for a bridge, as the bank was very low there – as is usually the case where a tributary, in this case the Fleet, enters a larger river. Because the river had to remain navigable, any bridge had to give boats at least as much clearance as London Bridge. Yet the higher its centre arch, the steeper the roadway up to the crown of the bridge would be, and horse-drawn traffic is badly slowed by gradients. And of course the lower the bank on which the approach road is built, the steeper the rise to the crown of the bridge.

There is no reason to suppose that Mylne had made a special study of bridges. He had certainly studied the ancient aqueducts of Rome, which are like bridges in reverse, carrying watercourses across a dry landscape. We also know of his study of the dome of St Peter's and the classical Greek temples of Sicily, and his interest in enrichments, ornaments and even life-drawing. He had also spent some time making plans of the ancient baths of Titus and Caracalla for another proposed publication.[7] Yet so far as he had shown any inclination towards specialisation, it was that of designing houses.

When he wrote in August, Robert asked for more money from the family. It was necessary, he wrote, that 'besides an ingenious drawing, an impudent face & good friends I should have a shell and Lodgings that bespeaks Affluence; For they won't probably trust a man with £160,000 that appears poor. Therefore my appearance before the Committee and others who are concerned, must be such as will give them no unfavourable idea of me.' The lodgings could wait till Christmas, but for clothing he needed 'a new shell from top to toe'. William should ask their father to make arrangements for him to draw funds in London immediately. Then two months would tell whether or not he was to be 'the greatest man in England in my way'.[8]

Later that month he wrote again, but to his mother this time and in a letter that gives a strong impression of his ebullient manner, as well as the way he larded his demands with snippets of gossip for her, and all in a homely Edinburgh style far removed from his usual correspondence. Something about it suggests that beneath the banter he was writing to his equal. The David Mylne mentioned was an elderly great-uncle.

Dear Mama,

I am always writing & always asking. To ask your pardon would be useless, unless I intended to ask no more. You that have something to give away to children who have nothing, must put up with inconveniences of that kind. My appetite for asking, I think, encreases as I turn older; But altho this preface indicates a demand of a very high quality, yet for the present I will be moderate and confine it to half a dozen shirts. A dozen would not be amiss, but I must be discreet with your wonted good nature. To tell you the truth I have very much occasion for them, a few rags that I bought in Italy serves me very badly for the present, & makes my face glow when I feel them rugged & ragged on my back in company. Remarks have been made to me that I must go genteeler; if I intend to Get good fish I must have the best bait, they say. Money has been offered me for that intent, but like a proud Scot I refused a boon, that would indicate the lack of a father & Mother. I expect my father's compliance to last month's demand that I may show them the contrary, & see if I can lay salt upon the taill of this Job of bridge building....

David Mylne is gone, God knows where. He has left off his business, gone out of England, and is upon some new scheme of life. It is all a secret. His wife who is a doudy body, won't so much as cheep it. I lodge with Mr Doby at present & find it convenient. If Bouls roll well tho this winter I must take a better lodging than anything they can afford. ... Remember me to all the auld wives & young lasses in your town end. And let me have the opportunity like a gratefull son to thank you for the boon I ask...

I am, Dear Mama, Your sincerely Affectionate Son,
Rob't Mylne[9]

A few days later he wrote to William again, through whom he now channelled requests to his father. He needed £20, or £15 immediately would do. Designs and estimates for the bridge had to be in by 4 October. Evidently his mother had already sent shirts even before receiving his request, and they had crossed in the post, so he thanked her. He said he was very busy, trying to show his merit to the 'Fat Aldermen of the City'. Meanwhile his bridge design was advancing, and was 'approved of by the principal man of the committee.'[10]

That was at the end of August 1759, and there is then a break of almost five months in the family correspondence. Fortunately events were by this time moving into a more public scrutiny, and there are records either published or recorded in the City archives to help fill the gaps. From these it becomes clear that Robert was right to think he had powerful support, yet this prize was not to be taken so easily as St Luke's.

When the appointed day arrived in October, the Committee found there was an application for a stay of execution. Two potential competitors were still finishing their designs, and as one of them was John Smeaton an adjournment was allowed, until 1 November. On that day it is said there were no less than sixty-nine designs put forward. The number of potential bridge-builders was slightly fewer, as some had put in more than one. Designs were entered both as plans and as models, and unfortunately almost none have survived. The competitors attended and were given an opportunity to answer questions. Some, no doubt, were dismissed with hardly a moment's thought, and eventually a shortlist of fourteen, including Mylne's, was chosen for further consideration. On 22 November the Committee met again, and now there was a heated discussion about the form of the arches in Mylne's design (see illustration 19-21).

There are many different types of arch, but the ones generally used in England at that time were based on a circle, either semi-circular or a smaller segment. It was a simple design to set out and very strong, and it was the one chosen at Westminster. To most people it was the obvious choice, and all entrants save one had favoured it. Mylne had instead chosen a flatter, elliptical form of arch. Some members present raised the objection that these were inherently weaker and less stable. These were not trivial objections, as everyone present will have known that bridge construction is fraught with difficulties, and all would have remembered that one of the arches at Westminster had collapsed soon after it was built, with a consequent three-year delay, unnecessary expense and unwanted embarrassment. It was not an objection that could be brushed aside, and the assorted merchants and tradesmen who formed the committee knew that they lacked the technical knowledge to decide it. The whole matter was put back so that eight 'gentlemen of the most approved knowledge in building geometry and mechanics' could be consulted.[11]

The trouble was that there were very few experts in the field. One of those chosen was Professor John Muller who taught the art of fortification and had written in support of elliptical arches in military structures, the others included the Astronomer Royal, a professor of mathematics who was known to disagree with Muller, a lawyer and a doctor.

There was then a delay of three months before the final decision, and during that time the whole debate spilled over into the public arena and became increasingly acrimonious. It is unfortunate that only one of Mylne's letters home during this period has survived, perhaps the only one he wrote, and that was towards the end so we do not have his private account of the early stages.

The first salvo took the form of an anonymous letter published in the *Gazetteer and New Daily Advertiser* on 1 December. Its author claimed that only three of the designs submitted for the bridge were being seriously considered, of which one had elliptical arches. He then went on to argue that such arches were inherently weak, and furthermore lacked what he called elevation and dignity. He then went on to hint that if such a design was chosen by the committee, 'what will the world believe than that some other motive than reason influenced the determination?' Worse, one of the experts to be consulted was 'Mr M-ll-r' who was known to be one who already shown a preference for elliptical arches. He then went on to flatter the bulk of the committee, calling them 'many of the most illustrious names of this great city,' and hoped they would not 'disgrace themselves and the metropolis of the kingdom' by supporting one of their number who 'instead of voting, aspires to dictate, perhaps without any claim to such superiority, either by greatness of birth, dignity of employment, extent of knowledge, or largeness of fortune.' The letter was followed by two more in a similar vein from the same source, each a week apart, and both pursuing the argument against what it called 'Mr M----'s design' with its elliptical arches.

So the attack was really two-pronged. First, elliptical arches were useless, and second some individual member of the committee, who lacked even such fine qualities as greatness of birth or largeness of fortune, was trying to force them through, for some 'other motive than reason'. Were the targets Mylne and Paterson and was the alleged motive their mutual Scottishness?

It helps to know who is the author of such letters, and fortunately it is known to have been Dr Samuel Johnson, a fact confirmed in Boswell's *Life* and elsewhere. He had written out of friendship for another of the competitors, John Gwynn. Five letters in all were published during December, three as we now know from Johnson, interspersed with two in which Mylne must have had a hand and of which he was probably the author. Johnson went on repeating the allegation of weakness, though modern commentators think he struggled with the science. He also questioned Mylne's qualifications. True, 'Mr M----' had won his architectural prize, but 'let it not be presumed that a prize granted at Rome implies an irresistible degree of skill. The competition is only between boys, and the prize given to excite laudable industry, not to reward consummate excellence'. In return came the defence that an elliptical arch, even if weaker, might still be quite strong enough for its purpose and more suited to this bridge, together with examples of its use at Florence and elsewhere in Europe.

No doubt the letters stirred up some coffee-house debate in the weeks leading up to Christmas, but more was in store. Early in January came the publication of a pamphlet that reopened the whole discussion. Written under the pseudonym *Publicus,* and cast in the form of an open letter to the committee it was called *Observations on Bridge Building and the several Plans offered for the New Bridge.* It ran to fifty pages and unlike Johnson's references to Mr M-ll-r and Mr M----, it spelt out the names of those it dealt with – Messrs Bernard, Chambers, Dance, Gwyn, Mylne, Phillips, Smeaton and Ware. It then set out and analysed the supposed merits and demerits of each of what were presumably the remaining serious contenders. It has to be said that this was no impartial analysis. The author's only real enthusiasm was for Mylne's design, and the whole book was directed to showing why it was the best.

It should also be said that despite its anonymity it was obvious to any reader that Mylne was either the author, or at least a major contributor. One only has to read it to see that the feigned impartiality with which it treats each contestant is tongue in cheek, and any lingering doubt is now dispelled by a letter he wrote to his father later the same month, where he said that '...in my defence, I have been oblidged to turn author in pamphlet and in Newspapers. I have been oblidged to speak in publick and reason with every species of men, from Astronomers down to porters.'[12]

Some, including Smeaton as we shall see, objected to this pamphlet, but looking back on it now it may seem unobjectionable. From the point of view of any individual competitor, each was faced with much the same problems in getting his design to the winning post. Each had used his own knowledge of the subject, whether limited or extensive, to produce what he believed to be the best solution to the conflicting demands. The number of arches and their type, the choice of materials, suitability for traffic of different kinds, navigability, strength, durability, beauty, cost, ornamentation and so on. All these things had to be balanced and choices made. Each competitor will have had reasons for his decision on each point. Yet the winning design would be chosen by elected councillors without any specialist knowledge of the subject, who were in truth unqualified to make the decision, and who might be unduly influenced by quite different motives, such as friendship, Scottishness, Englishness or a thousand other irrelevancies, so that the wrong choice might be made and the best design overlooked. It was very different from the usual kind of prize competition, like the one Mylne had entered at Rome, where the judges were practitioners in the field they were judging and needed no assistance from outsiders.

Against this background, why should an entrant not put forward the reasons his design should be preferred? Inevitably this would mean drawing attention to the ways in which it might be said to be better than its competitors, and so it could not be done until after submission day when all the other entries had been seen and could be assessed. Indeed a competitor who failed to do this and then saw his design rejected for what might be specious reasons would have only himself to blame for failing to expose the error. In this case matters had gone even further – Johnson had started the ball rolling with his anonymous public allegation that Mylne's arches were too weak, and that the committee would disgrace themselves if they chose it, and had gone on to repeat it week after week every time Mylne sought to rebut it. What else was Mylne to do? As he told his father, it was done in his own defence. Some might have tried to approach committee members privately to make allegations about the designs of others behind their backs. Mylne chose to set out his criticisms – and boasts – in the full light of day, and any of the others who wanted to do so had only to follow his example if they wished, in the weeks that remained before the final judging.

The *Publicus* pamphlet, of which a few examples have survived, is invaluable to an understanding of Mylne.[13] Even if he had some assistance in writing it, which is quite possible, it clearly reflects his own views and gives an insight into his priorities.

The text consists of a seven-page opening, followed by an average of four pages for each contestant in turn (Mylne got eight, all the others less), followed by a seven-page 'conclusion' that amounts to a further analysis of Mylne's design with some suggested amendments to it that the author somehow seemed to think he would consent to.

Taking the opening section first, it presented the committee with a stark choice. Would they do honour to themselves, their age and their country by erecting a strong, useful and magnificent bridge, or forfeit their reputations and render themselves justly contemptible in the eyes of posterity for many generations to come by leaving them a weak, inconvenient or mean one? The author claimed to be an individual of this city, who wished well to his country and to posterity. He analysed the problem of reconciling the competing interests of the men and carriages who would pass over the bridge, with those of the sailors who must pass under it, particularly when the river in times of flood reached the top of its banks at Blackfriars and boats would be floating at their highest through the arches. The bridge had to remain useable for barges and larger craft at such times, without having a roadway that was inconveniently steep for traffic. Its piers must obstruct the flux and reflux of the river no more than necessary for sufficient strength.

The opening section, only briefly summarised above, is also interesting in that the word posterity appears four times in seven pages, and was used again later in the booklet. As we shall see this was an important concept in Mylne's philosophy. There is no evidence that he had any strong religious beliefs, but he regarded future generations of humanity with respect, and always had them in mind when planning a piece of work.

After setting out the requirements of the ideal bridge, *Publicus* turned to each of the contestants in turn. He took them alphabetically, and conciliated in advance like a kindly schoolmaster with cane in hand, begging the pardon of the gentlemen concerned, and hoping they would realise that what came next was not against them, but had the sole purpose of serving the public and posterity. Then the cane came swishing down.

Mr Bernard, said *Publicus*, was an ingenious carpenter but this his first foray into stonework was not a success. He had copied architectural elements from elsewhere without understanding whether they were applicable here. He had crowded niches, pediments, plinths, pedestals and cornices on top of one another to support a recess like a bathing tub. His middle arch was high enough to give a very generous clearance for large river craft, but only by having those parts nearer the shore raised so high that the approach road would require a vast expenditure on earthworks and would be level with the second floor of riverside houses. Further, his ornamentation was both improper and mean, and that alone was sufficient to reject his design.

Next came Chambers, later Sir William and already architect to the Prince of Wales, who came from a family of Scottish merchants long settled in Sweden. He had studied under Piranesi, leaving Rome in the same year that the Mylnes arrived. *Publicus* began by conceding that his bridge had a grand, lofty and magnificent appearance at first sight. However, he seemed to have studied the architecture of the Romans without understanding how different was the society that gave it birth. He had the glorious idea of making a triumphal bridge, but in a manner that merely imitated a Roman style and showed no understanding of the very different customs of England. His bridge was thus crowded with urns along the sides and in niches – these were cremation urns, very suitable for the ashes of dead Roman heroes, but were English warriors in a country that did not practice cremation, stretched out cold after death in their country's service, to be crammed neck and heels into the same containers? Worse, his proposal to use brick for the piers and the infill of the spandrels was mean, and not of a piece with the rest. His cornice was trifling. He had used modillions, which Vitruvius told us imitate the ends of rafters – what place could roof members have on a bridge, and what sort of man would pick at a large stone until it resembled pieces of wood? Moreover, *Publicus* noted, moving from the sublime to the practical, Chambers had left a gap between the water stairs and the sides of the bridge that drunken pedestrians would soon find a use for, leaving a continual smell there to afflict those who went down to fetch water. His three middle arches were not only rather narrow, but a full 7ft too low for navigation. The piers varied widely for no obvious reason, had strange dimensions and were some too thick and others inexplicably thin. Their proportions were, he had to admit, beyond his comprehension, but they were obviously wrong. The shape of the cutwaters was also wrong, and quite unsuited to the tides of this river.

George Dance the elder was next, and as the City's own surveyor his entry would not be dismissed lightly. Also, Mylne had made a friend of his sons in Italy. *Publicus*, however, was no respecter of dignity. Mr Dance, he said, growing old in the city's service, might have learned a great deal from his long experience, but his design suggested otherwise. The arches nearest the shore were so high as that the banks would need to be raised 27ft – a monstrous expense, and so inconvenient as to be madness. It would mean an approach road at the same height, and what of the houses below, on either side of this supposed aerial street? Were they to have no access to the bridge? Or must they go inland nearly to the end of Fleet Market before the road was low enough for them to join it? It would make the most inconvenient entrance to a bridge that could possibly be contrived. His centre arches were fine, but if the outer ones were lowered sufficiently to allow convenient access, the roadway would be too steep for carriages. As to his ornaments, they were trifling, as was the cornice which he had managed to do worse even than Mr Chambers. His manner for laying foundations was quite impossible in a river where the tide rose and fell by 16ft. It would require piles 47ft long, a preposterous idea. His idea of hollow spandrels, proposed to save expense, would cost more in construction than it saved. The wooden centring he proposed for building the arches used twice too much timber and did not reach the standard of a country carpenter. His design was neither grand nor striking, and nothing about it could possibly outweigh its disadvantages.

Next came John Gwynn, later to be known as a designer of some fine bridges at Shrewsbury and elsewhere. *Publicus* thought his design had the attractions, and faults, of a Turkey carpet, generally pleasing but unsatisfactory on close examination. What may have been intended to be striking and magnificent had been turned into a trifling gewgaw. Although well drawn, the design was spoilt with some of its columns grouped like organ pipes, some surmounted by globes or pineapples, and others like dwarf columns or overgrown banisters. They were the ornaments of a pastry cook. His main arch was a little low and narrow, but like Mr Dance's his arches next to the land were so high as to need excessive earthworks, 21ft in this case. The design was certainly strong enough, but this seemed the only virtue *Publicus* could find.

Following his alphabetical sequence, *Publicus* came next to Mylne with an air that was respectfully lyrical. The design was 'magnificent, yet simple; light, yet not too slender; and modest, without bordering on rusticity.' It was 'noble and elevated … simple, and yet sublime'. It had 'no appendages but such as are genuine, and will stand the test of ages.' The cornice was 'grand and striking, and answers wonderfully well'. It was all 'a vast deal more than sufficiently strong'. Moreover, despite a main arch that gave a 28ft clearance above high water, the design only called for 12½ft of earthworks on the approaches. It was not all praise, though. The iron railing was trifling, though with a species of plausibility. The recesses along the footpaths, although necessary, did not project quite far enough. This, *Publicus* felt, was excusable, for 'when there is genius and attention to greater things, the lesser will often be treated with inaccuracy or negligence.'

He then went on to list and explain a string of ingenious technical innovations concerning caissons, arch centres, and ways of joining stones that Mylne had evidently struggled to explain verbally to the committee at the last hearing, one of them 'so simple that one is surprized it has not been found out before now', and all of which would save time and money, before adding disingenuously 'The author I hope will pardon me if I have forgot or misrepresented any of them.' In summary he thought Mylne's design 'a nice calculation betwixt strength and conveniency … grand and magnificent.'

Phillips, a carpenter, came next. He had put in no fewer than eleven plans, of which three remained on the table, which seemed to astonish *Publicus*. If one of them was better, what was the point of submitting the rest? Having examined them, he saw the only difference was a variety in the useless and ill-placed ornaments, though perhaps Phillips could feel easy on this score, as he had apparently not even drawn them himself. Perhaps a carpenter could transform himself into an architect with the help of a book – a *Palladio Londinensis* on his workbench – but it was not so easy to turn piles of stone and lime into the best form for a bridge across the Thames. His great arch was rather too little, his landward ends so high that 22ft of earthworks would be needed – the same problem as Mr Dance. His proposal to drive piles across the river from shore to shore and not merely where they might be needed under the piers was difficult to comment on. One of his designs had a building in the middle of the bridge, something like a temple or a summerhouse. *Publicus* suggested that if it was to be built, though quite contrary to the act of Parliament which specifically prohibited any buildings on the new bridge, it should have its name written on it to avoid misunderstanding just as, he claimed, the first painters of antiquity were wise enough, in their lame representations of nature to write under their works 'this is a cow' or 'this is a horse'. Further, although the toll on the bridge would only be a temporary one, Phillips had designed toll-houses apparently intended to last as long as the bridge would. His designs were the result of no great trouble, but rather a matter of drawing lines and twirling compasses without attention to numbers or quantity, a matter of copying and misapplying things ill understood. They need not delay the committee any longer.

Turning next to Mylne's most formidable competitor, Smeaton, *Publicus* surprisingly claimed that his design was the weakest of all those submitted. Whatever bridge was to be built must have one quality above all – any one of its arches should be able to stand alone in midstream if an accident damaged the next one – such things happened, as they knew from Westminster Bridge. The form of Smeaton's great arch, whatever one called its shape, was so wide as to cause a very great lateral pressure against its piers, and they were not strong enough to resist that pressure without the help of the rest of the bridge. He was relying on strong abutments at each end and arches that would all be built at the same time. If one arch was swept away, all the others must follow, like a game at ninepins or a house built of cards. When one considered that the thickness of his great arch was only 3½ft, it was even clearer what an absurd piece of mechanism it was. Judged as a piece of architecture, all who had seen it agreed it was the meanest and poorest of all the designs. He had lined it with what looked like a garden wall, topped with iron railings like those of private houses. He could not have found a worse shape of cornice for throwing rainwater off the sides if he had gone through *Palladio* from beginning to end. The stairs and the ends of his bridge were mean as well as awkward. Although his design would need little in the way of earthworks, and would provide an easy passage along the bridge, his arches were so low as a result that at high water only the very centre of the middle arch would be useable by boats and barges. *Publicus* even displayed the technical knowledge to accuse Smeaton of a draughtsman's sleight of hand: by not showing the high water mark on his drawing he had shown the bridge with an appearance which, if it were ever built, would only be visible for one hour every fortnight. Smeaton also proposed building his piers by the same impossible method as Mr Dance. It was the worst of all the designs.

Last and perhaps least came Mr Ware, whose design was partly good and partly bad. The middle arch was 33ft above high water, far more than necessary. As a result, like Mr Dance and others, he needed great earthworks on the banks if the roadway was not to be too steep – 23ft high in his case. As for his ornaments, although not very absurd there was no necessity for them and they could all be shaved off without doing prejudice to the work. It was lamentable to see the results of such jumbled ideas and want of method. Small inaccuracies might be overlooked, but when they related to the essential parts of the bridge it was impossible to prefer such designs without landing posterity – that word again – with something inconvenient for them and dishonourable for us.

Publicus reached his conclusion. He imagined he had addressed most of the important particulars. It followed that the designs of Messrs Bernard, Dance, Gwynn, Phillips and Ware could not be chosen because of the size of their earthworks. Neither could Chambers' because of the narrowness and lowness of his arches, nor Smeaton's because of the lowness and weakness of his arches. There remained only Mylne. His design gave an easy passage for men and carriages with little forced earth on the banks, and yet gave as good a passage under it for boats and barges. He had struck the just medium, thought *Publicus*, who seemed to have some inside knowledge of how Mylne's design might be modified if the committee insisted. Some had doubted the strength of his design, despite its evident superiority compared to all the others in utility, elegance, magnificence and the ingenuity of its construction. If this still troubled the committee, and *Publicus* urged that it should not, why then Mr Mylne's arches could easily be changed into the semi-circular form. He explained at length how this could be done, and added some new ideas showing how they could even be constructed so as to be infinitely superior to any bridge hitherto built. It would mean that the earthworks must be higher, of course, but only by 3ft making them 15½ft in all – far less than the competitors.

Finally, proposed *Publicus*, each member of the committee should put his hand on his breast and weigh the conflicting opinions over again, and if any tie of friendship inclined him away from the right decision, let him wrench it for a moment out, and, like Brutus, endeavour to do a service to his country and fellow-citizens.

If the claims of the pamphlet were true, Mylne should win. Any of the others who disagreed had only to reply, but only one of them did. *Mr Smeaton's Answer to the Misrepresentation of his Plan for Black-Friars Bridge* appeared later that month and sought to refute every one of the criticisms levelled at him, before ending with a challenge. The anonymous *Publicus* pamphlet had decried Smeaton's design in order to exalt Mylne's, 'but as Mr Smeaton can hardly persuade himself, that Mr Mylne himself could approve of treatment so unworthy of a gentleman, or an artist, he therefore leaves him with this alternative, either to publish the name of the author, or publicly disavow the whole performance; otherwise Mr Smeaton must look upon Mr Mylne to be himself the author of this ungenerous pamphlet'.

Mylne was never one to ignore a challenge, though his reply was not always in the form the challenger might have preferred. In this case it was *Publicus* who issued a three page *Postscript*, its pages numbered to follow on from the end of his *Observations on Bridge Building*. In it he repeated his challenges and deftly repudiated at least some of Smeaton's complaints. Of course he had described Smeaton's piers as 15ft thick and not twenty-five – they were only 25ft thick for two or three courses at the bottom, and then fifteen the rest of the way up. A pier would break at its smallest base, and it would not take the 25ft shoe with it as it overturned. If Smeaton had wanted his piers counted as 25ft, he should have made them taper to that thickness. It had given him pain, he claimed, to find fault with an individual, 'yet he, as an atom amongst the million of this city, must be set aside when the good of the whole depends upon it.' And he resisted the invitation to declare himself – 'Who I am it matters not' – before ending on a cheeky note. He could not but admire the earnestness with which Smeaton wanted to know the identity of someone who dared to find fault with his design, but if he really wanted to, then 'let him take the first man he meets in the street, and shew him his drawings.'

There was one more potential competitor, though he is not listed by *Publicus* and his entry seems never to have been submitted. This was none other than Robert Adam, settled in London and building up what was to become an enormous architectural practice, though Mylne had written home that some had already found him to be a charlatan.[14] The reason for believing that Adam planned to enter is the discovery of three of his drawings among the collection in the Soane Museum. They are dated to about 1759 and show a nine-arch bridge with a garlanded shield over the centre arch bearing the arms of the City of London; no other known bridge of that time would have contained those elements.[15] Why Adam did not make a formal entry after preparing the drawings is not known. He certainly harboured an odd resentment against the Mylnes, and it is possible that the risk of what he would see as a humiliating defeat at the hands of a younger man whom he probably regarded as a social inferior was a little too great. Once the result of the competition was known, the Adam brothers called Mylne 'Blackfriars' in their letters to one another, and his success seems to have rankled although they sometimes interacted professionally in the decades that followed.

The adjourned hearing eventually took place on 22 February 1760. By then replies, of a kind, had been received from the experts. Some had just said they knew too little on the subject to give any opinion. Muller supported Mylne's ellipses as predicted, while Simpson did not reject them. The matter was then put to the vote, and the waiting was over.

6

'... against the caballing interest...'

Once the committee's decision was announced, Mylne lost no time in writing home:

> Dear Father,
>
> It is with inexpressible pleasure, that I now acquaint you, my plan for the New Bridge in this city was approved of yesterday by a great Majority of the committee... Allow me to add with proper diffidence, an honour to my country and to those who brought me into this world and conducted me so far hitherto... The English Nation holds up its hands – and my countrymen stare. Nay, I cannot account for it myself – a young man just arrived in a great city, where he (k)new nobody, should, against the caballing interest of city factions, contrary to the interests of the Prince of Wales court, and spite of the specious plausibility of the Royal Society; steps at once into the head of his profession, and severall hundreds a year.[1]

The references to his competitors are intriguing. Chambers was architect to the Prince of Wales, while Smeaton was a prominent member of the Royal Society and had just won its highest medal for his research into wind and water as sources of mechanical power. The additional existence of caballing city factions confirms that Mylne did not have a clear run, and it is a pity that we do not know which other entrants were close to success or who their supporters were.

For a man sometimes thought arrogant, Robert's letters home show a pleasant humility around that time, and he wrote to his father often that year showing gratitude and affection in marked contrast to the firmness with which he had formerly claimed his delayed remittances. In January, while still awaiting the final decision on the bridge he wrote that nobody could have had a fairer opportunity of entering the world in the first rank of a profession. He was being talked of everywhere thanks to his bridge design, against the merit of which, he wrote 'there are no objections but those suggested by a reproach of my country,' and he asked for the family's prayers for one 'embarked upon the disturbed sea of this world.'[2]

The 'reproach' of Scotland was a reference to the anti-Scottish feeling that prevailed in many parts of English society and had been worsened by the current political success of the Earl of Bute, popularly believed to favour fellow Scots. In another letter Mylne told his father '...it is not a little that I suffered and do suffer for being a Scotsman ...'[3] That he was so targeted is confirmed by surviving literary efforts of varying quality – a street ballad, a satirical engraving and a piece of verse, all printed for sale.[3a]

The ballad takes the form of a rebus, a puzzle where pictures stand for some of the words or letters, done partly for fun and partly to make it more difficult for the target to sue for libel. Thus it has a drawing of a boot for Bute, an eye for the letter i, and so on. Two slightly different versions have survived, with different titles but almost identical verses. Substituting bracketed words for the little pictures, one is called *The (Bute) interest in the (City) or the (bridge) in the (hole)* (see illustration 18). Having categorised Scots in general as famous for *wh(oar)ing, cringing*, and suffering from the itch – scabies – it refers to Mylne's walk to Rome, complains *t(hat) the Scotch (shoe)d have power* and ends by complaining that even when the bridge is finished *twill (knot) be s(cott) free* – a pun on the fact that it was to be a toll bridge until the building costs were recovered. The other ballad, with almost identical words, has the different title: *(Bute) (awl) a new (ball) ad*, where Bute-awl is perhaps a pun linking the Earl's name to boodle, which a dictionary defines as 'money, especially for political bribery'.

Charles Churchill's poem *The Ghost* complains that a Scot should be building the bridge and not an Englishman:

> Englishmen, knowing not the guild, thought they might have the claim to build
> Till Paterson, as white as milk, as smooth as oil, as soft as silk,
> In solemn manner had decreed, that, on the other side of Tweed,
> Art, born and bred and fully grown, was with one Mylne, a man unknown.

As to the satirical engraving, two slightly different versions have survived. One has the title *Arriv'd from Italy – The Northern Comet with his Fiery Tail turned bridge builder Shews the artful section of his Stones*, and is a clear reference to the advantage Mylne had gained from his anonymous *Observations on Bridge Building*, from which it quotes. The other is almost identical but substitutes 'Puffing Phaenomenon' for 'Northern Comet' – a reference to the newspaper paragraphs that 'puffed' his success in Rome. In the coarse humour of the time it shows Mylne, labelled *The Phaenomenon*, squatting on a part-built pier of the new bridge, with his bare backside directed towards a group of well-dressed men in tailcoats and tricorne hats some with rolled plans – clearly the unsuccessful competitors. Mylne is fastidiously holding his own coat tails clear as he directs an effusion – in which can be discerned a medal labelled *Rome* and the words *observations on bridge building*, and *the fiery tail* – backwards at the group below (see illustration 17). As he does so he says 'Let them as I have done wrench it for a moment out and like Brutus endeavour to do a service to our country & fellow citizens' – the closing words of the anonymous pamphlet. They, in return make comments such as 'We shan't have time to clean ourselves before the decision – this is quantity with a vengeance', 'Did he learn these tricks at Rome?' 'Damme, is this your genuine, this your rigid utility, this the effects of his regular education?' 'I am not much bedaubed,' 'This is a strange mechanism,' and 'he gives us a specimen of Scotch presbyterian Eloquence & of Canny Edinburgh'. Separately, a man gestures towards Mylne and asks his woman companion whether she can see 'the Artfull Section of his Stones'.

On another pier of the bridge stands a group of Scotsmen. One, unlabelled but obviously Paterson, says 'I have secured him the committee', another says 'I told ye my good friend our puffing the Advertismets aforehand wad gain Credit with the Southerns – ye ken there is nothing like a good forecast'.

One of the defeated competitors also ventured into print with some gentle mockery – John Gwynn. In 1766 he published his suggestions for improving London with some observations on the state of the arts in Britain, and took a passing swipe at what must surely have been Mylne. All that was needed to turn a youth of modest parts into an architect, he said, was to apprentice him to a bricklayer, mason or carpenter so he would learn to draw a straight line. Then he should be sent to Rome, and after a time there should put a notice in the London papers to say that Mr Trowel the celebrated architect, on account of his vast abilities, has had prodigious honours conferred upon him, and shortly intends to revisit his native country, which he will no doubt do incredible honour. He can then return with a few drawings to put on his walls, and if some friend will beat his drum he will soon pass for a genius.[4]

None of these examples is particularly virulent, yet it probably irked Mylne to find himself a target for mockery when he believed himself to be performing a public service and when he had won the competition on merit, as the finished bridge would show.

Mylne's original design for the bridge was almost identical to the final version seen on the cover of this book, built in a gleaming white Portland stone. Considering first the view seen in profile from the river, the parapet made a long sweeping curve, rising and then falling to the opposite bank. From each bank the nine arches gradually increased in size, echoing the rise in the parapet until they met at the great central arch, which was wider and higher than those of the other two London bridges. Each of the piers between the arches had a pair of ionic columns which added greatly to the bridge's appearance while at the same time serving a useful purpose: each pair supported a projecting embrasure or recess at roadway level for pedestrians to pause and admire the view.

The original design had two other features never built. Instead of a continuous stone balustrade Mylne had chosen what was then a very innovative material

– cast-iron panels which could be made in one piece to fit between each pair of pedestals and would be much lighter than stone. These were intended to give users of the bridge a better view of the river and city, but were dropped in favour of a more conventional stone balustrade after objections.[5] Mylne had also wanted to put statues of British naval heroes on each pier, making it 'a repository of the actions of our admirals since England began first its present dominion of the sea'. This was a topic of great current importance as England repeatedly struggled with its European neighbours for maritime dominance, and London's prosperity depended on the freedom of the seas. The statues would be protected from above by the projecting embrasures and were designed to be viewed from the water, 'an element whereon these heroes have raised the character of this nation above all others and far above any one of antiquity' as *Publicus* put it. This idea was dropped during the building, probably because of the cost, as a total of sixteen pieces of statuary would have been needed to line both sides, though the effect might certainly have been magnificent.[6] Mylne's original drawing showing the cast iron and statues has not come to light. They can be seen, crudely drawn and wildly out of scale, in the satirical drawing *Arriv'd from Italy*, but with no way of knowing whether the artist copied them from a drawing he had seen or imagined them from the written description. The inaccuracy of the arch detailing suggests the latter (see illustrations 17 and 19).

For pedestrians and wheeled traffic the bridge presented a very different view. At each end was a broad new square, Chatham Place on the City side, Albion Place opposite, and the roadway and balustrades of the bridge widened in a sweeping curve as it reached dry land on each side, like the mouth of a trumpet as Mylne described it, gathering passengers from all directions as they approached and dispersing them again once they had crossed (see illustration 28).[7] For walkers passing over the bridge there was a pavement along each side, and the recesses already mentioned provided eight opportunities on each side for strollers to step aside and admire the changing view of city, river and sky without being jostled by the passing crowd.

At this time it was difficult for Londoners to reach the banks of the Thames, which were largely lined with buildings behind private wharves. A French visitor, who came in 1765 when the bridge was part built, wrote that Londoners had done their best to destroy or conceal the prospect of the river and the passages that led to it. He thought that the Thames might have been lined with palaces like the Grand Canal in Venice. Instead the banks were lined with the buildings of tanners, dyers and other manufacturers, approached by the dirtiest streets in the city. Even the bridges – London and Westminster – had massive parapets built as if designed to prevent any view of the river. These, he thought might be because of what he called the 'natural bent of the British, and in particular the people of London to suicide'. When he had gone in search of a river view, he had found it only by going into private premises.

What a contrast he found where the new bridge was building at Blackfriars. That bridge was to have only a rail for its parapet, at a suitable height to lean on. He did not name Mylne, but said he had spoken to its architect who hoped that the view from the bridge would encourage Londoners to line the banks with open quaysides behind which merchants and noblemen, and the richest of those who fitted out what the Frenchman called privateers, variously attracted by commerce or the noble view, would vie with each other to place fine buildings. To this end the architect was planning to start the line of a quay, so that others would simply have to follow its line 'when reason has overcome old prejudices'.

As for Mylne's bridge, he thought it would surpass even Westminster in its boldness and the magnificence of its decoration. He particularly liked the proposal for a pair of columns on each pier, supporting an elegant cornice that ran the whole length of the bridge, like the Pont Neuf in Paris.[8]

Turning from the matter of design to the progress of the work, on 27 February 1760 a week after announcing Mylne's success, the committee consolidated it by appointing him surveyor to the new bridge, at a salary of £400 a year to include £50 for a clerk. Twelve months earlier he had been eking out a student existence in Rome. Now he was a prosperous young professional, especially as it was understood that on satisfactory completion of the bridge his fees would be made up to the amount normal on major public projects, about five per cent of the total cost. The innocuous title of surveyor covered a whole variety of differing professional skills – Mylne was to be architect, engineer, surveyor, clerk of works, valuer and town planner for the surrounding districts on both sides of the river where the new approach roads would be built. Whereas a team had built Westminster, this was to be one man's bridge.

With this new status he needed a better address than cramped lodgings in Litchfield Street, and he found a home that provided him with plenty of space and a permanent view of the new works. The last house in Arundel Street off the Strand was at the water's edge by Arundel Stairs, on the north bank of the river about a quarter of a mile upstream from Blackfriars. As he wrote forty years later, he had given the task of building the bridge 'a continued and unremitting Attention, by Residence within View of that work for Twelve Years.'[9] Full of happiness with his success he found time to write some details to William, explaining that the move was '...partly to please my friends in the City and partly to be nearer my work. I have got a charming Appartment but it costs 50 guineas yearly. In Scotland I could get a whole mansion house and farm for the money.' At first he was there as a tenant, then moved into a larger apartment that became available. Later still he bought the lease of that and the other apartments in the building, partly with the help of a loan from his friend the anatomist Dr William Hunter, repaid when he received the final settlement for the bridge. A subsequent lease prepared when he eventually moved elsewhere and sublet it describes a very charming apartment indeed. Its finest room was 30ft long and twenty wide, wainscoted and plastered, with a chimneypiece of marble statuary and four windows looking out over the river.[10] It is easy to imagine the mixture of rapture and anxiety with which he must have watched year after year as his long-imagined bridge rose slowly out of the river in gleaming white stone. He also told William a little of his new domestic arrangements, 'my family and equipage consists of a clerk, a vale de chambre, a waterman and gondola.'[11] Even allowing for the fact that the waterman probably called his craft a wherry, this must have been a joyous time as the young architect, freshly brushed by his valet, stepped the few paces from his door to the bustle of Arundel Stairs to sit in the stern as his own waterman rowed him down to the bridgeworks like a sea captain being taken out to his ship.

Building the bridge is one of the few well-documented parts of Mylne's life. It was the biggest construction project of its time and progress reports found their way into newspapers, the *Annual Register* and the *Gentleman's Magazine*. Mylne had a clerk make copies of most of the written orders he gave to the contractors, as well as many of the working drawings, and these were later handsomely bound up in vellum as the Surveyor's Private Book. It includes precise drawings of each of the courses of masonry in the arches, stone by stone, as sent down to the quarries at Portland for the stones to be prepared. It is an invaluable record of how his major work was built and how he dealt with the problems that arose in the course of construction. It must have given Mylne a reliable record if any problems arose as work proceeded, yet there is an absence of the minor changes and scribbled calculations that are usually seen on working drawings, apart from a few explanatory notes in Mylne's hand, so it seems to be a master copy whose quality suggests that it was always meant to be kept. It passed down in his family who have since presented it and other papers to the British Architectural Library.[12] There is also information about the bridge in Mylne's letters to his father and brother, and he wrote some general descriptions in later years, as when he was petitioning the Common Council for final payment of his fees, and in a report on the whole project in 1784. Other sources of information once available have long disappeared, such as the minutes of the Bridge Committee up to 1768, which were destroyed when the tollhouse and committee room were set on fire during the Gordon riots in 1780. Nevertheless, enough has survived to provide a good outline, and two modern writers have made detailed assessments of the work.[13]

Before work could begin there were many preliminaries. First the precise location of the bridge had to be decided, as well as the angle of the crossing. To go straight across at right angles might seem obvious, but Blackfriars is just below a bend of the river. Mylne knew that it was important for navigation that the current should take boats easily through the arches with as little turbulence as possible, so he needed to know where the currents were and in which direction they flowed at different stages of the tide. This could only be done by experiment: he arranged for a 60ft timber spar to be floated into midstream and anchored at one end. Some miscreant cut this free in the night, whereupon Mylne replaced it with a larger one. By going out in a boat at different times he could observe the direction in which it floated, and thus decide that the angle of the current was between 69 and 70 degrees south of east in midstream.[14]

He also thought there should be wide straight approach roads – not usual in England – on both banks to make an impressive entry into the capital for foreigners arriving by road from the channel ports. This meant picking a line where straight roads could be built without demolishing important existing structures. The route he finally decided upon just avoided the end wall of Bridewell chapel on the City side and the churchyard of Christ Church on the Southwark side.

One problem he soon discovered was that the deepest part was not central but much closer to the Southwark bank. For best navigation, particularly at low water, he needed to put the largest arch of the bridge over the deepest part of the river. Was he to build a lopsided bridge with the highest arch, and the highest part of the roadway off centre? Further, there were extensive mudbanks on the City side where the Fleet debouched and unsavoury deposits accumulated. This meant that his landward arches on that side would all be standing in mud whenever the tide was out – unsatisfactory for a bridge whose designer was striving for magnificence. His solution was to propose new embankments on the City side, building the whole shoreline and some of the northern approach road outwards on built up ground over the mud. He could then set out the bridge symmetrically, and the central arch would be over the deepest channel, while the bridge at low tide would be seen standing in water and not mud except for a single arch at each end.

Once the precise line was settled, detailed planning could begin. The bridge would need vast quantities of materials – different types of stone for the masonry, and massive quantities of timber for various supporting purposes as well as piling. Suitable contractors also had to be found. In preparing their contracts Paterson's legal skills were deployed in combination with the practical knowledge Mylne had gained during his apprenticeship and later work as a young craftsman at Blair Castle. Contractors will sometimes quote a modest price for a job, buoyed up by the expectation that difficulties and changes of plan will inevitably crop up as the work is done, arguably not covered by the contract, open to negotiation and chargeable as an extra. Throughout Mylne's career there are well-documented examples of the disdain he felt for this practice, to which he would never give in on a client's behalf. At Blackfriars the contracts were drawn to try and exclude the slightest doubt as to how the work was to be done and paid for, with clear rules setting out the way work was to be measured. As a result one firm of masons and carpenters who had successfully tendered for the work demurred when they saw the contract they would have to sign. As Mylne later described it, they were therefore released from an agreement which, from the nature of their objections, they were 'not likely to execute with Spirit and Alacrity' and a considerable sum was saved by the use of such contracts. He noted with approval that the mason and carpenters who replaced them 'completely and honestly executed their Contracts.' The contracts included an entirely new feature – contractors had to quote two prices, one chargeable in time of war and one for peace. Most of the timber and stone for the bridge would be brought by sea, and when the bridge began the Seven Years War had forced up freight and other costs. The war was over and rates were down long before the bridge was finished, whereupon this precision in drafting was said by Mylne to have saved the City £5,839 – equivalent to two thirds of all his salary and fees for designing and building the bridge.[15] This kind of foresight, thoughtful but perfectly just, may have been one of the aspects of Mylne's approach that had made such an immediate impression on Paterson.

As the months went by, a question of mayoral vanity caused the foundation stone for the bridge to be laid in unusual circumstances. The Lord Mayor of London is elected for a single year and Sir Thomas Chitty, whose year it was, wanted his name on such a major structure. Plans were accordingly made for the ceremonial laying of a foundation stone on 31 October 1760 just before his year ended, but a slight problem was that none of the piers of the bridge had yet been started (see illustration 16). Mylne accordingly chose a piece of ground where the northern abutment was expected to stand in due course, and arranged for some stones to be laid there including one with a large cavity in it. The ceremony duly took place, with current coins of the realm being deposited as well as a plate with a Latin inscription. Before the Lord Mayor formally laid the stone that covered the cavity Mylne, in what was said to have been an impulsive and unplanned gesture, added his own contribution to the hoard. This was widely reported, for he was said to have added the medal he had won at Rome. As we know from his letter to William just after the presentation, the prize had consisted of two medals, not one. Even so it would have been a generous and dramatic offering to the fate that had carried him so far, and raises obvious questions. Was it something he intended to be found by distant posterity, or merely a symbolic offering to placate the gods of the river he was poised to conquer with his works? The former is more likely as the latter would suggest a superstitious side to his nature not hinted at elsewhere. Some may even have wondered if what he had placed there really was the genuine medal, for once the foundation stone was in place there would be no opportunity to inspect the deposit so long as the bridge stood. How long this might be none knew, but a properly built stone bridge will last almost indefinitely. London Bridge was six hundred years old, and it must have been expected that this bridge would also endure for centuries.

The City's decision that the new bridge should be built of stone brought its own problems, as there are no suitable quarries near London. Old Thomas Mylne became interested in the subject, and wrote to say that good stone might come cheapest from Scotland, but Robert probably felt that the project already had as much Scottish input as London would accept and the idea went no further.[16] Stone would have to come by water and Portland stone from Dorset was the obvious choice. It had been mined there since Roman times and although not very durable is a gleaming white limestone that would make a beautiful bridge. The first problem was that all the best quality Portland stone was in Government quarries controlled by the Treasury. One prospective contractor approached the Treasury for permission to use those quarries in June 1761, but their agent at Portland advised against it, saying it might cause supply problems to needful Government works, and so permission was refused.[17]

Following this a new application was made in January 1762, this time by Joseph Dixon, the mason who eventually took the contract and saw the bridge to completion. Mylne had by now located privately owned quarries on Portland that could supply suitable stone, and Dixon's application was simply to be allowed to use the King's road, pier and crane as needed to move and load the stone from the quarry to the coasting vessels that would take it to London.

The Treasury referred the matter to their agent at Weymouth, John Tucker, and his reply has the air of a jack-in-office peeved because Dixon had presumed to go over his head. There had been no necessity, he wrote, for troubling the Lords Commissioners of the Treasury with the matter, as it was well known that the free use of His Majesty's crane and the road to the pier was never refused. It was of course subject to paying the usual duty and following any necessary regulations, and might be delayed if it would cause any interruption to His Majesty's service, but it was simply a matter of applying to His Majesty's officer on Portland – who happened to be another Tucker, in this case Richard.[18]

The Treasury passed this news on to the bridge committee, but difficulties in getting stone out of Portland became one of the greatest problems Mylne had to cope with, just as it had earlier caused Wren endless delays at St Paul's. According to Boswell, this was because, despite the best efforts of the Bridge committee, 'Parliamentary interests ... often the bane of fair pursuits, thwarted their endeavours.'[19] The Scottish stone Thomas had proposed would have been cheaper, less trouble and probably more durable – but it would not have been a popular choice, and would not have had the lovely white colour of Portland.

Thomas had also encouraged Scottish masons to travel down and seek work on the bridge but Robert, having placed one thus sent 'with one of the best masters here', wrote asking him to send no more:

> My reason I hope you'll approve of – To receive an obligaton from such sort of people as Master Masons here are, may give them hopes of being requitted, at the expense of the good of the publick, my charge. Would you believe it, father, I have been drawn into the temptation of bribery, varnished over with every symptom of politeness, yet, thanks to you, I have had the virtue to withstand it & even spurn it with contempt.[20]

He clearly knew the danger of putting himself under an obligation to those who might be less scrupulous about what they expected in return. It is a pity he was not more specific about the bribery. Was it bribery whereby he would be paid for favouring a particular supplier, or over-certifying the amount of work done, or was it that Mylne was expected to bribe some functionary in order to get smooth access to the materials he needed? Whichever it was he obviously knew the public purse would be the loser if he complied.

Mylne paid at least one visit to the quarries himself in mid-March 1762. His diary makes no mention of it, but George Lucy who he had met in Rome was staying in Bath and happens to mention in a letter that he had spent some time with Mylne, who he said was visiting Bath after a trip to Portland to choose stone for the bridge. They spent a couple of days together, and Mylne was able to bring him up to date with news from Rome.[21]

Mylne's notes show that every course of stone for the bridge piers was the subject of a drawing showing the shape and position of individual blocks, and copies of the drawings were sent to Portland so that each stone could be shaped in the quarry by masons there – a sensible procedure when blocks weighing several hundredweight apiece were needed. The correspondence with the Treasury did not end the supply difficulties – there were also labour problems of 'combinations' of workmen, such that Mylne had to write despairingly to the Bridge Committee in August 1763. It is clear from his surveyor's book that stone sometimes took more than a year to arrive after it was ordered. The committee wanted the bridge built in seven years, but two and a half had already gone and he had barely been able so far to get 5,000 tons of stone delivered. As he needed 36,000 tons in all, the bridge would take eighteen years to finish at that rate, and he asked the committee to use their influence in the matter, as he had done everything that was in his power. Later, in 1766, the quarrymen refused to work, and when masons were sent down from London to cut the stone the locals starved them out by denying them access to food, drink or lodging.[22]

Even when the labour difficulties were overcome and the cut stone had been laboriously hauled to the pier, weather could still play a hand; Mylne recalled years later how often adverse wind and weather added to what was already a slow process, for the distance by water was over two hundred miles. On one occasion it took sixteen months for stone to arrive after it was ordered.

As to its suitability, Portland stone is relatively soft, and parts of the bridge exposed to the impact of carelessly handled barges or the effects of ice soon showed the scars. Ice was much more common then, partly because the river was wider and shallower at the edges, partly because the damming effect of London Bridge slowed the current and gave ice the time to form. On 30 December 1762 Mylne noted in his diary that the river was caked with ice, and all work had to be suspended until it cleared on 8 February. Occasionally the ice on the Thames was thick enough for Frost Fairs to be held, when crowds thronged the river and oxen were roasted on the ice, which happened as late as 1814, but none of these very severe winters happened while the bridge was being built, though there was still enough ice and frost for the masonry to suffer damage on many occasions. Below the water line, by contrast, it endured very well despite the constant currents. In 1833 an engineer was sent underwater with the help of 'Mr Deane's patent diving helmet' to examine the state of the piers, and was able to report that all the stone below low-water mark was as good as the day it was laid, with the mortar entire in its joints and the tool marks still clear on the faces of the stones.[23]

Not all the stonework of the bridge had to be from Portland. Purbeck stone was also used and Kentish rag for some less important parts. The rag stone came from quarries near Maidstone, and was taken on barges down the Medway and thence into the Thames without any sea passage, making for a much easier journey. Other parts of the structure simply needed to be packed with gravel, and that was easily dredged from the river-bed. When the bridge was eventually demolished it was noted that those parts contained fragments of clay tobacco pipes mixed with the gravel, and a walk on the foreshore in central London at low tide will discover some of these even today – an odd reminder that such pipes were used and disposed of in vast quantities, having a very short life not only because of their fragility but because the clay absorbed the tarry tobacco juices and soon made them foul-tasting.[24]

Designing the elegant superstructure of a bridge is one thing, turning it into a real structure with its foundations firm in a swirling riverbed is another, and all the more so when it has to be built to a tight budget. England had many stone bridges dating from medieval times, but few bridge-building skills had survived into the eighteenth century, as Nicholas Hawksmoor soon discovered when he was consulted about the proposed bridge at Westminster in 1736 and had to refer to continental texts because there was nothing suitable in English. The Thames presented special difficulties. Many continental rivers dry to a trickle in summer, exposing most of the bed and greatly simplifying the task of building foundations. Many also run through bedrock, an ideal foundation. By contrast, although the Thames at Blackfriars rose and fell by up to 16ft, it did so twice a day as the tides flowed, and the bed was never revealed except in shallows near the shore. The bed consisted of mud overlying many feet of gravel that had washed downstream over a long period, with the original chalk riverbed far below it. The upper layer of gravel was very loose, and Mylne noted that the boats that dredged it up for ballast did not have to move as they worked. They dropped anchor and scooped it up, and as fast as they raised gravel, more flowed in to replace what they had taken. Anyone who has struggled to cross even a dry shingle beach will understand the problem. He had to find a way of putting immoveable foundations on this slippery bed.

The method used at Westminster had been less than satisfactory. Labelye, the Swiss bridge-builder, had originally considered driving piles to provide a firm base, but wrongly decided it was unnecessary after making test borings. When his bridge was almost complete one of the piers tilted and eventually settled more than 5ft, so that the arches that joined it to its adjacent piers had to be hurriedly demolished until it could be rebuilt, for fear that a collapse would take the whole bridge with it. Labelye's bridge had thirteen arches, inevitably making it look less elegant than Mylne's sweeping nine.

Mylne chose to use a combination of known methods, each refined by him to meet local conditions. The basic method was to provide a solid foundation by driving rows of wooden piles into the riverbed at the site of each of the piers. Stone piers would then be built on top of the piles, and these piers would in their turn become the foundations for the arches that completed the bridge and carried the roadway overhead. Each of these stages carried its own problems, but Mylne showed himself able to produce innovative, and usually economical, solutions to each problem as it arose.

The first stage of the work involved driving piles. As well as a stone abutment on each shore, eight intermediate stone piers had to be built across the river to carry the nine arches, and tests soon showed that at each location the bed was loose gravel, so piles were needed to provide a firm foundation. Oak would have been the traditional choice but in the course of his test borings Mylne had found the remains of an ancient jetty made of fir, which was still in excellent condition and he decided that Baltic fir would be the timber to use. Time proved that he was right, and it is an example of how experimental the whole project was and how adaptable he was prepared to be. When specimens were cut from it in 1833 it was seen to be in a sound and perfect state, 'as bright and fresh-coloured as new timber'.[25]

Because the lower part of most of the piers would be permanently under water, he had to find a way of building them in dry conditions. In some rivers it is possible to build cofferdams to isolate a piece of the riverbed, which can then be pumped dry to allow work to proceed. This was impossible at Blackfriars because the bed could not be waterproofed – the gravel was as porous as a sieve, and more water would well up through it as fast as it was pumped away. The remaining option, which he chose, was a caisson – a floating box that could be weighted down, sunk at the right spot, and then used as a dry working space, although the floor of the caisson would of course prevent any direct contact with the riverbed.

First the location of a pier was chosen, then guard works were built. These consisted of an outer rectangle of temporary piles, driven into the river bed and extending up some 25ft above it, so that they would protect the works from collision by passing vessels at any stage of the tide. One section was removable to allow the caisson to be floated into the enclosure in due course. Around the perimeter a vertical wall of planking was driven into the bed so that when gravel inside was excavated it was not immediately replaced by more flowing in. Loose gravel inside this perimeter was then scooped out, sometimes to a depth of several feet, until a firm layer of gravel was reached. All this work had to be done in swirling tidal water.

The next stage was to provide a firm foundation by driving rows of wooden piles into the gravel as far as the ground would allow, which varied from about 6 to 12ft. It had to be a strong foundation, as Mylne calculated the total amount of Portland stone needed for the bridge at 36,000 tons, in addition to which there were large quantities of infill and ballast. Such a load spread over eight piers and two abutments suggests that over 4,000 tons rested on each of the bases.

The machine for this task was adapted from a design used at Westminster, where some piles had been driven. It was mounted on a moored barge, and worked by the traditional method of hauling a heavy weight – the ram or 'monkey' – to the top of a wooden framework, from where it could fall many feet to land like a hammer-blow on the top of the pile – on one occasion Mylne noted that it made the nearby ground shake. Then it would be hauled up again and the whole process repeated as often as necessary. Mylne increased the weight of the ram to 14-hundredweight and this could be raised 30ft by horses walking an endless circle on the barge to turn a capstan that raised the load.[26] All this entailed much labour. Each of the eight piers needed between eighty and a hundred permanent piles, quite apart from all those that had to be driven and later removed for the guard works.

Variations in the riverbed meant that some piles could not be driven in as far as others, yet Mylne needed all their tops level in order to make a platform for the bases of his piers. This he achieved by modifying another machine that had been used at Westminster – a saw connected to a framework with ropes and pulleys that could be used underwater by two men standing in a barge anchored above (see illustration 26). This was used to trim the tops of all the piles to the same level, and he ordered the carpenters to build it so that it would work in 14ft of water and cut within an inch of the riverbed. Using a saw horizontally is never easy because gravity hinders the making of a clean cut and tends to jam the blade; to do so underwater by means of ropes and pulleys must have been very hard work.[27]

Once the grid of level piles was ready inside its surrounding guard works, the next task was to build a pier of strongly jointed masonry there. The base of each pier was about 85ft long and 33ft at the widest. Mylne's solution was elaborate and involved building a caisson to enable work in deep water. The use of a floating wooden caisson is simple in theory, but became complicated in this case because of its size. It had to be high enough for the top of its sides to be above the highest tide when its bottom was resting in an excavation several feet below the natural bed of the river, and this made it the height of a three-storey building. It had to be rigid enough to remain waterproof under massive pressure yet buoyant enough to be manoeuvred across the river like a boat. To fit round the future pier it had to be wider than it was high and about three times as long – something like the shape of a cardboard shoebox. Such a structure would float much too high in the water to be stable without ballast. When built, it was 86ft long, 33ft wide and 27ft high. Caissons had been used at Westminster but those had only been 16ft high, with the disadvantage that they could only be worked in at low tide, having to be pumped out each time the tide fell before the masons could set to work.

To give an idea of the size of the caisson, ten double-decker London Routemaster buses could have been parked side by side in it and still have left room for their drivers to stroll around the perimeter. A second fleet of ten could then have been placed on top of the first, yet passengers on their top decks would still not have been able to see over the sides.

Such a structure was subject to colossal forces whether floating or submerged, from the combination of wind, tide, current and water pressure. Of these only water pressure was predictable, depending only on the depth of submersion. Wind was unpredictable. The speed of the current varied between inshore and mid-stream and also according to the state of the tide and the flow of the river – after a period of heavy rainfall in the Thames Valley the level could rise several feet and increase the speed with which it flowed down to the sea. As for the amount of tidal rise and fall, that cannot be predicted with certainty even today, as it depends not only on the phase of the moon but also on the amount of water in the river and the direction and strength of the wind, which can help push water out or pen it back.

The rise and fall of the tide was potentially Mylne's enemy, but one he turned to his advantage as others have done, by using it as a gigantic hydraulic lift that could at least be relied on to rise and fall twice a day. He devised a three-stage plan for building the caisson. Its construction began close inshore, just below high-water mark. It was then floated out to a lower level, below low-water mark where its structure was completed. Finally it was floated out further still, to the deep-water site where the pier was to be built.

First his men drove a grid of piles as big as the planned caisson's base into the mudflats at the river's edge, and attached to them a heavy wooden deck. This made a kind of vast fixed workbench on which the bottom of the caisson could be built. It was just below the level of ordinary high water, so that work could continue without difficulty at most stages of the tide. Work began in June 1760, on the mud bank opposite the mouth of the Fleet River as all the other nearby shallows belonged to the owners of wharves who had the valuable right of mooring their craft anywhere along their frontage down to low-water mark. Meanwhile a second fixed platform of the same size was being built in deeper water nearby.

The caisson bottom, built of massive timbers like a huge raft, was 85ft 4in long by 33ft wide and 2ft 3in thick.[28] It was built on the first platform and progressively ballasted with stone as its size increased to stop it floating away when the tide came in. When it was complete some of the stone was unloaded so that it could then be floated off on the next high tide, and anchored above the second platform, on to which it sank when the tide fell. It was then loaded with more stone taken out by barge to keep it from floating away on the next tide. Work then began to build its wooden sides, which were 1ft 6in thick and 27ft high. The lower parts could only be built when the tide was out, but once the sides were high enough work could be continuous. Mylne calculated that the finished caisson weighed about two hundred and fifty tons, of which two hundred was the Riga fir, twenty the iron fixings and thirty was masons' and other gear (see illustrations 23–25). Floating was the only method by which such a vast structure could be moved, and Mylne's system meant that the river did much of the work for him.[29]

Inside the caisson masons then began to build the bottom course of the stone pier Mylne had designed, which when complete made a mass of carefully jointed stone about 2ft 6in high filling the whole floor of the caisson. That much stone weighed about 400 tons, but the caisson was still buoyant enough to float on a high tide, and it was now time to move the whole structure to its final position. The stone had a dual purpose at this stage, first to provide the ballast necessary to keep the caisson manageable when afloat, and then to become the bottom layer of one of the piers of the bridge. Mylne then picked a suitable moment for the whole caisson to be floated with the help of barges to a position inside the guard works and over the level area of sawn pile tops that had been prepared to receive it.

Pumps were needed to keep it afloat, as there was inevitably some leakage although it was designed to be waterproof. There was a temporary, and leaky, joint between bottom and sides, because there was no way of removing the bottom from under the hundreds of tons of stone that would be built on top of it. Instead it would remain in place, with the tops of the piles below it and the superstructure of the bridge on top of it, while the sides of the caisson could be removed and attached to a new bottom.

Once the caisson was inside the guard works, the masons continued their work, 25ft below water level at times but protected from the river by the caisson sides, until four courses had been laid. By now the stonework was 10ft high and weighed over a thousand tons, the caisson settling ever lower in the water as barges took more stone out to it, but still afloat at each high tide. It was then time for the sides to be taken away as there were structurally-vital internal braces that prevented working any higher inside the caisson, while the top of the masonry was by now above low water mark, so the masons could work on it whenever the tide was out without needing the protection of the caisson. Of course the whole structure had to be floating in precisely the right position over the piles at this point. Careful measurements were taken as the tide went out, and when Mylne thought right, a sluice was opened to flood the interior so that the whole structure settled on the riverbed. This was not an exact process, and sometimes the caisson had to be pumped out and refloated to make some adjustment to its position. The joints between the bottom and sides of the caisson were then undone. During these works the caisson often needed to be pumped dry and here Mylne is said to have been the first to use an innovative kind of chain pump, which was thereafter adapted for the use of the Navy.[30]

The guard works were then removed at one end, and two barges floated out to be moored one on each side of the caisson. At low water they were firmly attached to it by a framework of wooden beams with triangular braces, and as they rose with the tide they lifted the caisson sides. The lower part of one end of the caisson was hinged, and this flap was then raised, leaving that end of the caisson high enough to clear the top of the stonework. The barges then took it back to the deep-water platform to be joined to a new bottom, which would in due course become the foundation level of the next pier. This method was ingenious and probably the best that could be devised. It was economical because only the bottom of the caisson was lost each time a pier was built while the other four sides could be reused. Even the bottom was not wasted, as it became the base of one pier of the bridge, which would otherwise have needed a layer of masonry of the same thickness.

One contemporary drawing that shows a problem with the caisson is to be found in a variant of the *Northern Comet* satire. It is later, and has an extra scene added below the main picture. It portrays an event in May 1761 when the caisson was to be launched for the first time but the tide did not rise high enough to let it float from the fixed part of the stage, so that 'the populace were disappointed of part of their entertainment.'[31] The picture shows the broad expanse of the Thames with Westminster Bridge in the distance and no sign of the new bridge except the vast caisson, which is firmly aground on its stage. Under a crescent moon symbolising the tides, Mylne sprawls exhausted on the riverside mud calling to a man opposite 'My good friend of Monkwell street quick lend me your hand to help me out of this sh-tt-n affair for I am fallen in my own Filth, oh my stars what a condition shall I be in if I am to be here till next new moon'. The other man is plainly Paterson, who used Monkwell Street as his address on letters, perhaps because his livery company, the barber surgeons, had their hall there. Paterson, his legs knee-deep into the holes of what seems to be a double-seated privy, replies 'The deel tak your Mud Box you silly loon, I am plung'd as deep in your filth as you are and want to gett out myself.' The picture includes some detail apart from the caisson – Mylne has his sword at his side, and by his hand a speaking-trumpet that he presumably used to direct the work.

The Surveyor's Private Book contains precise information about the way the stonework of the piers was built. Each course of masonry was 2ft or 2½ft high, and the stones were cut to a complex plan with many of them slightly tapered so that they would not work loose. In addition, once they were in place a double dovetail, shaped like a bow-tie but 20in long and cut from oak 5in thick, was set in sockets cut across each of the internal joints to hold them together even more immovably. This was a method he had learnt from ancient greek temples he examined in Sicily.[31a]

One of the disadvantages of such work was that the bridge had to be built under the continuous scrutiny of watermen, passing citizens and the merely curious. Many wished the project well but there were certainly others who would have enjoyed the young Scotsman's discomfort when things went wrong, as they inevitably did, and others still who must have pestered him with the benefit of their advice. Because of the uncertainty of tidal heights the caisson sometimes began to float when it had been thought securely anchored, and at other times it had to be left part-loaded for months awaiting the arrival of the next shipment from Portland. On one such occasion in late 1764 ice formed in the river and the caisson had to remain sunk until the following April. By then it had deteriorated and could not be pumped dry, and for weeks Mylne must have suffered agonies of humiliation before work was back on course. No doubt the bridge committee sometimes took it out on him when their friends, and the friends of the defeated competitors, became a little too mocking in their enquiries as to the health of the bridge works. Again and again he tried more effective ways of pumping, until there were six horses and eighty men on the caisson working nine pumps of different types, but each time the river defeated him. The first such time he elegantly noted in his diary that he had eventually 'desisted', but later entries just say that he 'gave it up', until success eventually came in late May.

By now the caisson was in use for its sixth pier, had suffered years of weathering on the river and being a simple wooden box whose joints were subject to enormous forces was nearing the end of its working life. After being used for one more pier it was dismantled and its long sides became the bases of the landward piers on each side, where shallow water simplified the process of making foundations. Mylne's design for the caisson had proved itself, its work was complete and all the inevitable problems had been overcome.

As soon as the first two piers had been built, work began on the next phase, building the arch to link them. The first piers built were the central ones, so the great arch was the first to be built, which had a span of 100ft. A stone arch cannot support itself until the last stone is in place, and a temporary support of centring is needed until that moment. Mylne used wooden centring to support the arch, but with two useful innovations (see illustration 27). He had designed the lower parts of his piers so that they thickened as they approached the riverbed, each course being stepped out wider than the one above it. This made for a larger base and thus spread the weight of the bridge over a larger area of riverbed, but it also provided a series of firm ledges on which he could rest the ends of the struts of his wooden centring. This was labour-saving by comparison with Labelye's method at Westminster, where he had driven piles on which to rest his centring and later had to saw them off – two unnecessary operations, and wasteful of timber. Mylne's method also meant that because the centring rested on the base of the pier, the pier gradually took up the weight of the arch as each new stone was added. Labelye's system by contrast put the weight on the piles, and the pier only received it abruptly at the moment when the centres were removed. Further, Mylne had designed his centres so that they could be taken apart and reused for other arches, a considerable saving of materials and labour. He had also designed them so that only three sets of centring were needed for the whole bridge, and this meant a vast saving of timber.

All the controversy there had been about his elliptical arches must have heightened the anxiety as the first arch took shape and reached the stage where its centring would be removed. As the arch was built its weight bore down on the centring which supported it, pressing the centring ever more firmly against the bases of the piers on each side. By the time the arch was complete and the centring became redundant, this loading would jam the centring in place and make it very difficult to remove. Mylne devised an elaborate system whereby rows of multiple oak wedges set between copper sheets were inserted under the centring before any stone was laid, raising it all a few inches. Once the masonry was finished these crocket wedges could be gradually eased out, which allowed the centring to drop a little so that it was no longer jammed tight against the underside of the arch and could easily be dismantled. Years later Mylne claimed in a letter that when the wedges were removed as he stood under the keystone of the 100ft arch 'the hand that writes you this trembled not.'[32] Perhaps not, but his diary shows the very careful note he took of the state of the new arches as they settled into position, with some inevitable movement as some joints opened or closed a little, piers settled and minor cracks formed. All the arches were successfully built in this way, and none ever gave real cause for concern.

Next after piers and arches came the roadway of the bridge, added only when the main structure had had time to settle a little, as it inevitably did. This was designed to form a smooth curve along its length, rising to the crown where it levelled and then fell gently again to the far side. This was an improvement on Westminster, where two straight slopes met in a sharp hump at the crown of the bridge.

By 1766 the bridge was built apart from the last two arches on the City side, and it became clear that the small expense of a temporary structure to complete the bridge for toll-paying pedestrians would be worthwhile. Mylne designed a wooden bridge to serve this purpose, which opened in November 1766 and charged a halfpenny toll, so that the bridge was thereafter in use for pedestrians and was beginning to pay its way just over six years after construction began.

For one group of people the bridge was bad news. London's watermen constantly ferried people and goods backwards and forwards across the river as well as along it. The crossing trade would disappear from that part of the city as soon as the bridge was in use, and the Act for building the bridge contained a provision that the watermen must be compensated. A few years earlier in 1758 when two central arches at London Bridge had been taken down to be replaced with a single large one more convenient for boats, a temporary wooden bridge made to plug the gap as work proceeded had been destroyed in a mysterious fire. When Mylne planned his temporary footbridge there were fears for its safety, as there had already been some malicious damage to the bridge works. At around that time he started negotiations with the Rulers of the Watermen on behalf of the bridge committee, and on 19 September 1766 agreed the substantial sum of £13,650 in three per cent stock as a one-off compensation for their loss of trade. Oddly, just four days later, one Webber who had been sentenced to transportation at Maidstone Assizes for a shipboard robbery, and who for some reason wanted to be hanged instead, made a long confession to a string of other crimes including the 1758 arson at London Bridge. Whether that was true or false, on 28 September the Rulers of the Watermen gave public notice of the generous settlement they had reached with the city and said that if any watermen should do any mischief to the temporary bridge, they would cooperate with the city in seeing the offenders brought to justice.[33]

There were two more civic ceremonies after the laying of the first foundation stone in 1760. The first was on 23 June 1761 when Sir Robert Ladbroke and others of the bridge committee went on board the caisson to lay what was described as the first stone of the first pier, with an inscribed slab of black marble to commemorate the occasion.[34] The second was on 1 October 1764, when the first of the controversial elliptical arches was complete – the great arch in midstream. Mylne had kept the waterway blocked with floating booms, and on that day he opened the booms to let the Lord Mayor and Sheriffs 'be the first that ever went through' in their ceremonial barge, after which they went ashore for a grand celebration (see illustration 22). When the whole bridge was completed and opened to wheeled traffic, in November 1769, there does not seem to have been any ceremonial. It was originally to be called the Pitt Bridge after the Prime Minister, but Blackfriars was a more obvious and less partisan choice, and that was the name that stuck. Completion of the bridge was not completion of the project, as much work still remained on the improvements to the embankment on the City side as well as the approach roads.

On the City side Mylne laid out the wide square of Chatham Place leading into what was then called Bridge Street, now New Bridge Street, to meet Fleet Street at what is now Ludgate Circus. On the south bank Albion Place led into a straight new road that ran to a new junction, St George's Circus, where it met the roads from Kent and Surrey. Now Blackfriars Road, it was originally called Surrey Street and led to an obelisk designed by Mylne still to be seen at the Circus. This sets out the distances to Fleet Street, Westminster and London Bridge, each about one mile away. Surrey Street was built across marshy ground by tipping more than 70,000 cartloads of rubbish, and was accomplished in little more than a year, so that it was in use by 1771. Mylne had wanted to make an even greater improvement to London traffic by running two straight new roads starting from the south side of the bridge, one east to London Bridge and one west to Westminster Bridge. This would have allowed traffic between the City and Westminster, where the courts were situated as well as Parliament, to take a direct route south of the river avoiding the badly congested route along Ludgate Hill, Fleet Street, the Strand and Whitehall. Unfortunately there were too many competing private interests, each seeking its own advantage, so that agreement could never be reached and the opportunity was lost for ever.

On the City side there was a great deal of riverside to be embanked. This work took many years to complete, all under Mylne's supervision. He later gave evidence to Parliament of the endless negotiations it had required, especially with the societies of Inner and Middle Temple, who gained acres of extra land in the process.

The cost of building the bridge was to be paid for by a toll on all users, but the bridge was so popular that most of the cost was recovered within sixteen years. On 22 June 1785 the bridge was made free to all traffic except on Sundays, when a reduced toll of a halfpenny was levied on foot passengers to pay for the cost of lighting the bridge, and this continued until 1811.

By the time the bridge was open Mylne had proved himself and was known everywhere. The bridge was a major benefit to London. It adorned the river, eased congestion on the other bridges, saved travelling time and provided wonderful views of the city. It had been partly built in wartime and the supply difficulties had been considerable, yet it was ready in a reasonable time and was built for almost exactly the estimated cost. The net cost was just £152,840, compared to £218,810 for Westminster Bridge. All the responsibility of construction had been his, whereas Labelye at Westminster had been one of a team with shared responsibilities. London now had three river crossings, of which Mylne's was the finest looking, and no more would be built in his lifetime. Critics were kind to the design, then and since. Sir John Soane thought it would 'long remain a monument to the artist-like mind of the architect. Its lightness, fanciful originality and picturesque appearance will always command the approbation of those who consider architecture as a pleasing as well as a useful art.' Summerson called it 'an excellent piece of work'.[35]

It was probably Mylne's greatest achievement, and could have remained for centuries with occasional repairs. Sadly the narrowness of London Bridge meant that its nineteen tide-constricting arches were removed after a new five-arch London Bridge by John Rennie had been built alongside in 1823–31. The effect was a 5ft drop in low water level above London Bridge, and according to Smiles the unimpeded velocity of the tide quickly began to scour the river deeper until the foundations were exposed at both Blackfriars and Westminster, and that combined with the vast growth in traffic meant that each in turn had to be replaced.[36] In the case of Blackfriars, a new bridge by Joseph Cubitt, of five arches faced with cast-iron on granite piers, was ready in 1869, but even that had to be widened by half within forty years. Such was the pace of London's growth that Mylne's beautiful structure had lasted for just a hundred years.

7

'... on the pinnacle of slippery fortune...'

As he celebrated his first New Year's Day in London in 1760, optimistic about the bridge competition but with its outcome still unknown, Robert Mylne must have wondered what the future held. It was not for him to know that he had five clear decades ahead, but looking back now one can see how neatly those decades separate themselves, each marking a different phase of his life and each characterised by its own distinct mixture of problems and solutions, pleasures and sadness. The first four, taking him to the end of the eighteenth century, can each be looked at in a single chapter. The last part of his life is more complex and will merit a longer look.

Building the bridge neatly filled the sixties, with the first stone laid in 1760 and the first carriages crossing in November 1769, though the linked roads and embankments took years to finish. Yet he had time for other matters, and his family was one of them. One letter he wrote to his ageing father in 1760 ends with a striking phrase rooted in classical sources that reveals a reflective side to his character – 'Adieu ... may you enjoy the retirement you are in, and when you have a leisure moment, pray for one who stands on the pinnacle of slippery fortune and the world's esteem.'[1A] It seems he was not too dazzled to sense the risks ahead.

He was also concerned about his unmarried younger sisters Jean and Anne. Two years earlier their older sister Elizabeth had made a runaway marriage, eloping from the house at Halkerston's Wynd where she acted as William's housekeeper, and Robert was anxious the others should not be tempted to follow suit, by ensuring they had proper dowries. Writing to his father he urged him to make the best provision he could, and offered to contribute a part of it, though his sisters need not know that he had done so: 'it shall appear as if from you – our whole fortune in this world is owing to the manner of stepping into life. They, their children and children's children will bless you for it when we shall be no more ...'[1B]

All this time his mother was in good health, and he kept her supplied with small consignments of luxuries like tea and sugar, but his father was failing and had handed his business over to William. As early as 1758 while still in Rome, Robert had written to William with more than a hint that Thomas enjoyed life to

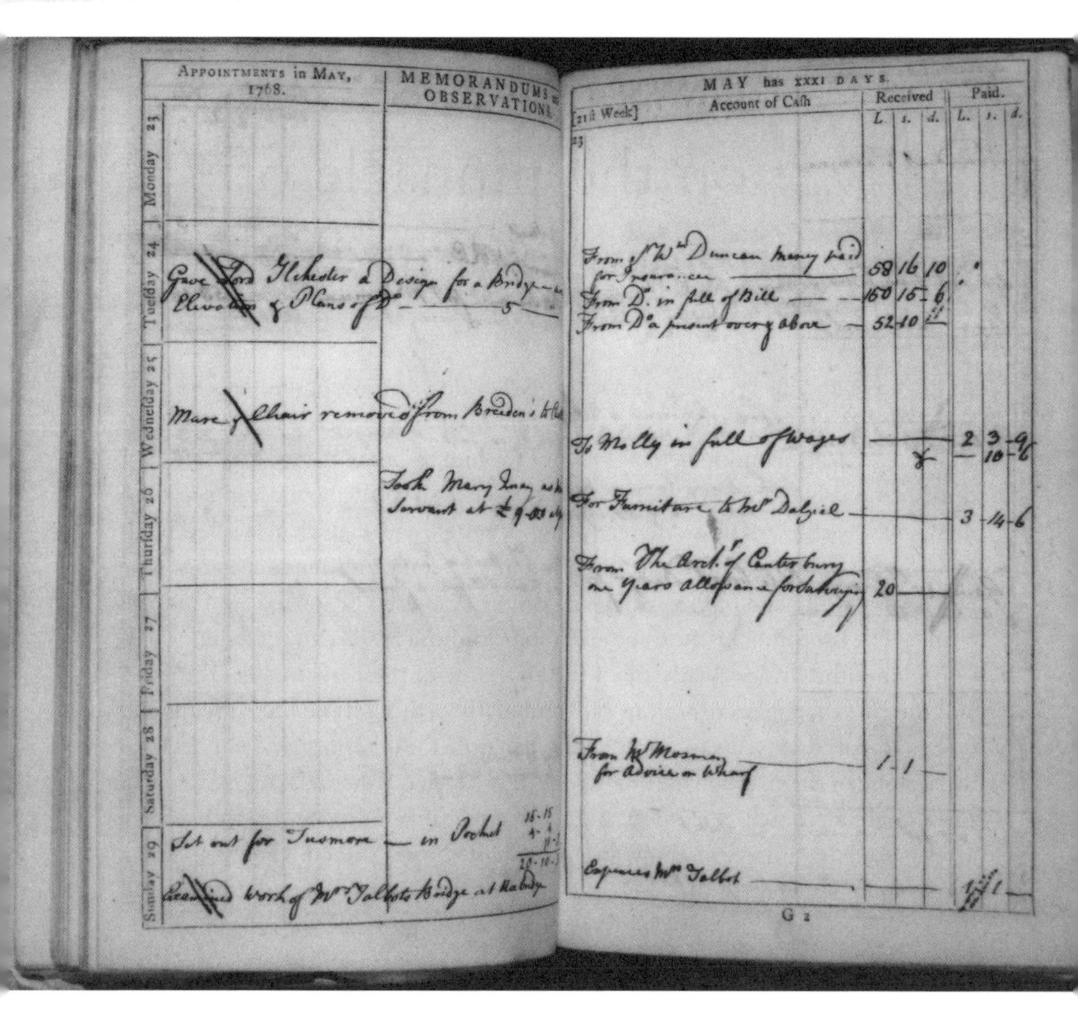

Appointments in May, 1768.	Memorandums & Observations
Monday 23	
Tuesday 24 — Gave Lord Ilchester a Design for a Bridge – an Elevation & Plans of Do. — 5	
Wednesday 25 — Mare & Chaise removed from Breeden's to [illegible]	
Thursday 26	Took Mary [illegible] as Servant at £9 [illegible]
Friday 27	
Saturday 28	
Sunday 29 — Set out for Tusmore — in Pocket [illegible]. Examined work of Mrs Talbot's Bridge at Uxbridge	

MAY has XXXI DAYS.

[21st Week] Account of Cash	Received L.	s.	d.	Paid L.	s.	d.
23						
From Sir Wm Duncan money paid for Insurance	58	16	10			
From Do. in full of Bill	150	15	6			
From Do a present over & above	52	10				
To Molly in full of Wages				2	3	9
					10	6
For Furniture to Mr Dalziel				3	14	6
From The Archbp. of Canterbury one years allowance [illegible]	20					
From Mr Mosman for Advice on Wharf	1	1				
Expenses Mrs Talbot				[illegible]		

G 2

Robert Mylne's diary, week of Monday 23 May 1768, showing the printed layout and a typical mixture of business and household entries. The left-hand page shows design work for Lord Ilchester; the removal of his mare and chaise to new livery stables; the arrival of a new servant earning £9 a year; a note of cash at the start of a journey to Tusmore where he was building a house for a client; and a reminder that he had examined Mrs Talbot's bridge at Uxbridge.

The right-hand page lists sums received from Sir William Duncan, the Archbishop of Canterbury and Mr Mosman, as well as the final wages and gratuity to a departing servant and other minor expenses.

Most of his entries were not as legible as these. (RIBA, British Architectural Library, drawings collection) the full '...You have alarmed me not a little with his indisposition, for I hope he refrains from that excess of society which renders a man in our country 80 years in his constitution before he is 40; let this be graven in your heart ...'[2]

In February 1763 Thomas wrote Robert a letter, now lost, that must have called him home, and Robert who was going through a difficult stage with the bridge works and had just parted company with his clerk replied promising to come as soon as he could. He found a new clerk and set out for Edinburgh within four days, but arrived to learn his father had died a few hours earlier.[3] They buried him that week, and Robert stayed on for another week during which he was made a Burgess of the city. He and William dealt with the estate, and Robert as the elder inherited the crumbling mansion of Powderhall, where his widowed mother and sisters continued to live, while William stayed on in Halkerston's Wynd, which they still owned as well as other Edinburgh properties.

In 1762 Mylne began to keep a diary, and anyone wanting to understand him is likely to feel a surge of interest on hearing that he did so from the age of twenty-nine until a few months before his death forty-nine years later, and that the diaries still exist.[4] Disappointment swiftly follows, for they do not tell us much about events or people, nor did he write of his feelings, tastes, preferences, opinions, hopes or disappointments. What he recorded for the most part were those events in his working life that would need to be billed and might otherwise be forgotten or misremembered. This makes them more like account books than journals, and indeed this is what they were sold as, most having titles such as *Kearsley's Gentleman and Tradesman's Pocket Ledger*. Of course, it tells us something of his life to learn that on a given day he gave Lord Morton some drawings for the design of a book-room, or dined with Sir William Duncan and gave him advice worth half a guinea, but it makes dull reading and takes us no closer to an understanding of what it would have been like to have been there. Was it at home or in a tavern or club? What did they eat and drink? Who else was there and what did they talk about? If, like Pepys or Boswell, he was ever tempted by a housemaid or afflicted with a hangover it was not something he felt the need to record.

Although they spread over half a century, the diaries are almost identical, and resemble modern pocket diaries – printed for a particular year, and with an identifiable space for every day. They were about 6in by 4in, small enough to fit a large pocket, and originally bound in red leather. As well as diary pages they contained general information which varied from year to year, such as lists of MPs, peers and bankers, postage rates, journey distances, stage coach routes and costs, points of law – anything that might tempt a wavering buyer. The diary part always had fifty-two double-page openings, one for each week. All the left-hand pages had seven ruled spaces for the days starting on the Monday, and all the right-hand pages had ruled columns for cash paid and received, a division Mylne usually adhered to. He did not make entries every day, and there are blanks for weeks and sometimes months on end. The handwriting is necessarily tiny. The most regular entries on the left-hand pages are those that note work done and to be charged for, and these remain legible despite being struck through with one or two diagonal lines, presumably when a bill was sent or payment received. Money actually paid or received is on the right-hand pages, always with a person's name and often a description of what it is. These financial entries are usually even duller than those opposite, yet they are sometimes the only source of a particular fact – for example, that he at one stage had a country retreat near Croydon is only disclosed by notes of the rent paid and the cost of furniture, gardening and housekeeping there.

As well as work done Mylne often noted the date and wages when a new manservant or coachman started. This was necessary because wages, typically £12 to £15 a year with free board, were not dispensed on a regular weekly or monthly basis, and could only be calculated accurately if dates were recorded. Servants often left or were dismissed, and this was recorded with a note of the final settlement. Thus we learn quite a lot about household affairs, though the motive for the entries may have been purely practical. He also noted the handful of occasions when he took on a new salaried post, such as surveyor to St Paul's, and again it may have been to keep track of the salary calculation. For most of the diary period Mylne was married, and a few of his diary notes suggest that his wife kept a separate series of housekeeping books, but these have not survived.

Apart from these mundane entries Mylne noted such things as the towns where he spent a night while travelling, and some of his appointments, but very rarely who he had met on a given day unless a bill was needed.

As to his overall financial position, in the first year or two he tried to keep a running total of money in and out. This then stopped, probably because he realised the total was meaningless, as it included funds he handled on behalf of others, such as the Bridge committee, or the Scots hospital. He obviously had to record these but they were never in any sense his money. As well as his own personal income and expenditure he also needed to keep track of such things as travelling expenses. He was meticulous about this, carefully splitting the total between two or more clients visited on the same trip out of London, with a deduction for his own portion if he had made some personal detour. The care with which this continued to be done over half a century speaks highly of his integrity. Very occasionally he went back over a year's entries and prepared a summary of his finances on a plain sheet of paper. A century or so after his death one of his descendants had the diaries rebound, perhaps because they were starting to deteriorate. At that time they were grouped five years at a time and these few loose papers bound in among them. Taken together with other financial information in the diaries they give a fuller account of Mylne's growing prosperity than we are ever likely to have of his emotional or social life. It does not follow that he regarded material wealth as pre-eminent; rather it was something that a meticulous professional had to keep track of. There are also indications from jotted additions to some entries that he wanted his children to learn from his example the benefits of prudent personal economy.

In the first of the diaries is a printed exhortation from the publisher: 'As these Books may be considered as the Annals of a Man's Life, and may be of Use, even after his Decease, they ought by all Means to be preserved, to have Recourse to, when Occasion requires, to consult or prove, any Receipt or Payment, Memorandum or Observation, Time or Occurrence.' It is doubtful that Mylne needed any such injunction to preserve books that he knew to contain valuable information, and it is fortunate for us that his descendants realised their historic value and have made them available to researchers.

Of his personal life there is very little in the diaries, except where some entry casts an indirect light on his lifestyle, such as the annual tax on carriages, the cost of having his post-chaise regilded, or the regular purchases of wines and spirits. As to family matters the pattern is the same – we learn about the boys' education only from the cost of their school fees and the girls' from the occasional note that a language teacher or governess has been engaged. Dinner guests are almost never mentioned; a frustrating omission as we know from other sources that Dr Johnson and James Boswell were among them, and a record of such contacts spanning half a century would have been fascinating. Family deaths, and some births, are recorded but without comment. One explanation for this may be that he needed to give his clerks access to the diaries to ensure that the fees he noted were charged to the clients. If that were the case, the diaries would never have been a place for private reflection. It may also explain why the handful of annual financial summaries he made was on loose pieces of paper although there were plenty of suitable blank pages in the diaries.

A selection from these diaries was published fifty years ago but was later realised to be what Colvin the architectural historian called grossly inaccurate. Perhaps there was some muddle in the editing, but the inaccuracy can be rapidly confirmed from the originals. To take an example from the first year, Mylne is quoted as having written the phrase 'cast and wrot', in the course of an entry setting out measurements of one of the bridge piers at Blackfriars. That has the look of an archaic technicality – is it perhaps about some use of cast iron and wrought iron, for we know that Mylne occasionally used the two materials together, as at the staircase of Whitefoord House in Edinburgh, or did he mean that he 'cast' the measurements in the sense of totalling them, before he wrot(e) them down? It is the kind of oddity that may have given many readers a pause for thought. Anyone who checks the original will probably decide the phrase reads not 'cast and wrot', but 'east and west', which makes perfect sense in the context.[5] Later the same year, and this is not an exhaustive list, meanings are changed as when 'cut' becomes 'put', 'fenced' also somehow becomes 'put', while ordinary 'neap tides' become the mysterious 'ninp tides'. The final entry that year, still dealing with the state of the river, changes Mylne's account of a 'great' frost that filled the river with 'cakes' of ice, to a 'grave' one that filled it with 'layers' of ice – almost the same meaning, but not Mylne's words. Furthermore, the entries printed come exclusively from the left-hand pages, and even then are mysteriously selective, ignoring perhaps a third of the total. Having chosen to include the entry describing the ice-caked Thames on 30 December and that it caused all work on the bridge to be suspended, why omit the entry five weeks later that records 'Frost ceased and ice cleared away from works'? Omitting all the right-hand page entries is puzzling, as they are sometimes the only source of an interesting fact, such as that Mylne once commissioned a portrait from Sir Joshua Reynolds and also bought a Canaletto. As long as such things are borne in mind the published diary, which begins with a biographical essay, has its uses and must have involved considerable labour, but potential readers deserve to know its limitations, especially as it nowhere indicates that it is only a selection from the diaries.[6]

So Mylne's diaries are little help in telling us how he spoke or thought. One rare description comes from James Boswell who arrived in London in 1762 after dropping out of Glasgow University, where his father had moved him from Edinburgh's fleshy temptations, with vague thoughts of entering the catholic priesthood. His father, the judge Lord Auchinleck, must have seen this as a move from frying-pan to fire, and relented so far as to arrange an introduction to the wealthy Lord Eglinton, a Scottish friend living in London who had a circle of 'high-born and rakish friends including the young Duke of York, brother of the future George III'.[7] It was at dinner at Eglinton's that Boswell met Mylne, and made a note of the conversation, which turned to the difference between rough and polished societies:

> Eglinton said that a savage had as much pleasure in eating his rude meals and hearing the rough notes of the bagpipe as a man in polished society had in the most elegant entertainment and in hearing the finest music. Mr Mylne very justly observed that to judge of their happiness we must have the decision of a being superior to them both, who should feel the pleasure of each; and in that case it would be found that although each had his taste fully gratified, yet that the civilized man, having his taste more refined and susceptible of higher enjoyment, must be acknowledged to have the greatest happiness.[8]

This single opinion confirms what we might expect, that Mylne put a high value on the civilizing effects of cultured pleasures, but how much better we might understand him if he had noted such thoughts himself from time to time.

Mylne's increasingly comfortable lifestyle during the 1760s can easily be seen from the right-hand diary pages. Thus in 1765, as well as furniture, linen, cutlery and plate, he also bought 'a picture of Canaletti's', presumably a small one, for £3 15*s* 0*d*. The subject is not recorded, but Canaletto lived and worked in London from 1746 to 1756 and painted many views of the Thames, while Mylne had some knowledge of the picture market from Rome, where he had advised visitors like Lord Garlies on buying paintings.[9]

There are some entries for personal items, so we learn that his best clothes included a lace-trimmed suit and waistcoat, that he had a sword, a horse and chaise, and later had his own coachman, dressed in livery and with an allowance for boots and buckskin breeches. Household servants came and went, and there was a succession of Sarah, Mary, Ann, Molly, Jean and Jane, variously described as undermaid, servant, cook or housekeeper, usually two at a time and paid £7 to £9 a year as well as their board and some tea and sugar – valuable commodities. When they left they sometimes received 'and a little more' beyond their strict entitlement and the turnover, averaging about a year's service, may have been normal for London. He sometimes offered a rising scale that increased every year to reward longer service. He had a waterman, Isaac Parsons, who must have constantly shuttled him from Arundel Stairs down to the bridge works and back again: virtually all the work on the bridge was accessible only by river. Parsons had £30 a year, but that presumably included the use of his boat. Mylne had what he described as a clerk, who must also have been a draughtsman. They too came and went, and were well paid, at between £30 and £50 a year depending on whether or not they were lodged in the house. One from this time, Thomas Cooley, later distinguished himself as an architect in Dublin, and working for Mylne must have ranked as a good training.

The diaries probably give a fuller picture of his growing professional practice than any other aspect of his life. He clearly tried to jot down a note when he carried out some chargeable work, whether in the form of drawings, reports, valuations or verbal advice. He also sometimes noted the reason if there was to be no charge, as when the client was an old friend or sometimes for works of public benefit. This would hardly have been to aid his own memory, so possibly his clerk made out bills from the diaries and needed to know what was exempt. Routine work for employers who paid him a fixed salary was usually not recorded, and Blackfriars Bridge is the only one of his works where he diarised many of the details and problems of its construction, perhaps because it was such a major work that he wanted to leave a permanent record of all the problems overcome.

His other work in the 1760s was pleasingly varied. Private clients, including Dukes and many other peers, often but not always with a Scottish link, as well as an Archbishop, several Bishops and many other prominent or wealthy people in England, Scotland and Wales. Some he had met in Rome, others were recommended by existing clients. There is no obvious example of his seeking architectural work, but it is clear from the letters he wrote to his brother from Rome that he expected future commissions from many of those he met there, and his growing reputation probably brought all the work he wanted. The diaries show him taking on a good deal of architectural work during the 1760s, and he probably expected it to become his main activity. He travelled all over the country designing new houses, and altering others to modern tastes, from grand country houses like Tusmore (see illustration 1) and Wormleybury to gothic follies and ornamental bridges. His London work was equally varied. For William Almack he built assembly rooms in King Street, designed for weekly subscription balls and the home of a mixed club where new applicants had to be elected by members of the opposite sex. For his friend Dr William Hunter he rebuilt a house in Great Windmill Street to provide an anatomy theatre as well as a library and private museum, said to be of great size and with admired proportions for the doctor's collections, which included one of coins said to be second only to that of the King of France. He also designed more modest private houses, often for acquaintances. He surveyed cathedrals and castles. In Edinburgh he designed St Cecilia's Hall, an oval concert room based on an Italian model and still in use though much changed, which his brother then built, and a town house for Sir John Whitefoord.[10] Apart from purely architectural work, he was also in demand for advice on water matters – bridges, harbours and docks, and in due course canal work when it became available. The Duke of Portland asked for a bridge design for Welbeck Abbey, and wanted a very large single arch. Mylne agreed, and went regularly to check how the workmen were proceeding, but the quality of stone available locally was not what it might have been. As the months went by there were reports of stone falling away and eventually the arch collapsed, something that Mylne later said had given him much occasion for self-reproach, and possibly his only significant structural failure.

One of Mylne's earliest houses was at Cally in Galloway, for the agricultural improver James Murray of Broughton, a relative of Lord Garlies who he had

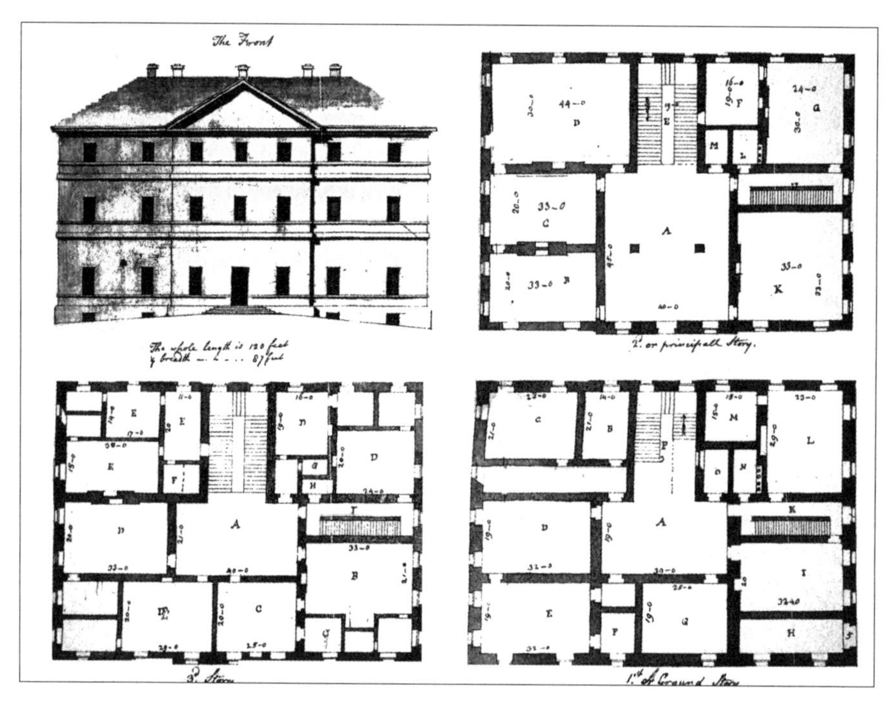

Preliminary elevation and floor plans for James Murray of Broughton. These are Mylne's earliest surviving house plans, drawn in Rome in 1759 for a house to be built in Galloway, Scotland. They accompanied a letter in which he set out in detail his philosophy of house design. (National Archives of Scotland)

met in Rome. While still in Rome he sent preliminary proposals for that, and the letter happens to survive. Although the design was later modified once the client's requirements were better known, the letter is useful as showing the way he approached the problems of house design. Of course the house must present a handsome appearance, but practical considerations were always in the forefront of his mind. Accompanying a set of small plans and elevations, his thoughts are set out in full, with key letters to match the drawings and it is striking how his musings bring the imaginary house to life:

> The situation is supposed to be upon a small rising ground, on the easy declivity of a hill, fronting south south-east. An extended plain before, for prospect: and the hill arising behind to preserve it from the cold winds. This house is divided into 4 stories. The first is for the use of all the servants who work in the house. It holds the kitchen and all the nauseous places that should not be seen or smelt by company. It is half sunk in the ground to keep it cooll and half above to give it light and make a pedestall to the whole building. The entrance to it is by the private stair at one end to keep the front free of the drudgery servants. Below it are the cellars for coolness. The second storey is entirely appropriated for the reception & entertainment of Strangers. In this story the whole day is spent. The 3d Story is all Bedchambers for the Master, Company, & children of the house. The 4th Story is rooms for the principall servants, nursery & a few Rooms for more company. The Garrets are for the lower servants who work in the house. The roof is Bevelled all round and has leads on the top. Round the eaving of the roof are spouts to collect the rainwater into a cistern in the 4th Story, for the preservation of the house against fire, and by pipes may be conveyed to the different Storeys for the use of the Chambermaids.
>
> In the 1st or ground story, A is a hall which gives admittance to all the rooms on this levell ... it is lightened from the stair and must be painted very white. B is the larder, which as it should be cooll, is turned to the North. C is the servants hall, placed in the corner, that their noise may be out of hearing. At the end of the passage which leads into these last two Rooms, the knives are cleaned. D is the housekeeper's store room, which as the housekeeper is often called in time of dinner, it is placed alongside the second table room. G is the Butler's Pantry. H is an easy descent for the hogsheads down to the cellars, which come in by the great door on purpose. I is the Steward's counting room, it lyes next the private stair & entry, and has a door from thence for to admit those people who have business with him. K is the private stair which goes with one flight from this floor to the next above, and with another down from this floor to the cellars, at the end of it is the door for the entry of servants, tenants &c. L is the Kitchen, which as it should be clean & cooll, is turned to the morning sun and coolness of the north. It is placed under the library to keep its disagreeable smell from the rooms where company are. M is the coall house, by the window of which

the coalls may be thrown in from the carts. N is the office house for this understorey, lightned from the kitchen. O is the scullery, lightned from the Great stair P. All this Ground Story must be vaulted for strength & fear against fire – In the 2d or principall Story, A is the great hall, from which there is admittance to every room without going through one another. In the middle of it are 2 square pillars to support a wall on the 3d story. B is the small Dining parlour, which as it is principally made use of for winter & small companys, it has its windows to the west & south to enjoy the sun at the time of dinner. C is the drawing room & is placed here for company to retire to either from the parlour or great dining room. Its windows look west to enjoy the sun at the time of day that this room is made use of. D is the grand dining room, which is so placed that if it is too hot weather, it may be made cooll by shutting the western windows. It is 5 feet higher in the ceiling than any room on this floor. E is the great Stair, open from top to bottom & to the halls. It must be made of wood all hanging. F is the Masters dressing room, it has its entrance from the great stair before it goes down to the first story. For commodiousness it is placed along side of the Study or Library G, which has its windows to the East and North, to have the morning sun for study and for business, & these winds for the preservation of books. It is placed above the kitchen that it may be drier, it has its entry from the private stair for the easier reception of the tenants. H has this commodiousness of a private stair, that with 3 flights of steps, it brings you up to the 3d floor, and the 4th flight is a levell passage which gives admission to the library. K is the finest room in the house for its ornaments. It is to receive visits in, breakfast and dance in. L is the little house for this story, & lightned from the library. M is the coall house for this story, & is enlightned from the stair.

In the 3d story, A is the hall which gives admittance to all the rooms on this floor .... B is the principall bed chamber for the Master and Lady of the house, it has its windows east for to enjoy the early sun. On one side it has a closet and little house, & recess for the head of the bed. On the other it may have a door to the private stair. C is the Lady's Bedchamber, turned to the South ... alongside is a closet for its use. DDDD are 4 Excessive good Bedchambers for company, with closets to each.[11]

It is the writing of an architect who believes his primary task is to design a machine for living in, a house that caters for all its occupants' complex needs for views, heat, light, safety, grandeur, solitude, display, privacy, quietness, and practicality, with rooms placed where the day will be at its best when they are most in use. Perhaps all good architects have these things in mind but it is reassuring to see them so clearly set out. The drawings show that the exterior is as elegant as it should be, but it does not rate a mention in the letter, save that the basement will act as a pedestal for the upper parts and that the roof will be designed to harvest the rain for household use. It seems that his clients' needs would come before the demands for exterior ornament or unrelenting symmetry.

In addition to his growing architectural practice Mylne showed a willingness to take on other duties. When he first saw London in 1759, St Paul's Cathedral, completed less than fifty years earlier, was the most conspicuous and for many the loveliest of its buildings. He could admire its dome and west front from Arundel Street, and the view from the new bridge was even better. When in 1766 he learnt that the cathedral surveyor had died he lost no time in writing to one of his most influential clients, the Duke of Portland, to seek his help. He told him of the vacancy and, mistakenly, that it was in the gift of the Dean of St Paul's to whom Mylne had already written. As the current Dean was Bishop of Salisbury and a brother of the Archbishop of York Mylne pointed out that the Duke knew the family, and asked for the goodness of his recommendation. He added in an anxious postscript, 'There are several Candidates and the Decision will be made soon, as the man died a fortnight agoe.'[12] A few weeks later on 7 October he had the obvious pleasure of writing in his diary, 'Appointed Surveyor of St Paul's Cathedral by the Archbishop of Canterbury'. Whatever the Dean's views or the Duke's efforts, it was indeed the Archbishop of Canterbury who made the appointment, and he later noted in his memoirs that appointing Mr Mylne 'the architect of the New Bridge' to the post had been at the request of the Lord Mayor, the third trustee in addition to himself and the Bishop of London, so perhaps Mylne had also been seeking support in the City.[13] There is a caricature of him by Nathaniel Dance-Holland, newly appointed and seeking more funds for the fabric from Archbishop Secker (see illustration 6).

Why he wanted this job is a matter of conjecture. It was hardly for the salary, which at £50 a year would make little difference to his lifestyle. This is borne out by Farington's diary note after Mylne's death almost fifty years later that it was 'a place of Honour, but the salary only £70 a year.' Was it merely for personal status, and the honour it would bring to his family name to have charge of such a prominent building? Or was it because his admiration for Wren's masterpiece had grown as he gazed from his vantage point on the bridge, and the idea of having it within his care was irresistible? It can certainly be shown that he had a great veneration for Wren later in his life, and he may have been smitten from the moment he saw St Paul's with its surrounding thicket of Wren churches gleaming palely in the sun. This last explanation seems the likeliest, for according to James Elmes who knew him in later years, 'his love and affection for ... St Paul's ... was such that he never would see it defaced, or altered or spoiled in any way, and scarcely a week ... passed without him giving it a personal survey.'[14]

The post at St Paul's was one that he held until his death. The likelihood that he did not seek it for the salary is confirmed by an earlier occasion where he sought the Duke of Portland's support as well as making his own direct application. The position of Master of Works in Scotland had become vacant, and Mylne must have thought it would make him a substitute for the long defunct post of King's Master Mason held by so many of his ancestors, and would also give him the chance to examine some of his forebears' work. When he heard that Hamilton of Bargeny, a Member of Parliament, was also asking for the job but would lose his seat if he got it, he suggested to the Duke that if he could have the title and duties, he would undertake to pay the salary to Hamilton. That way, as he pointed out, Hamilton could keep his seat, and:

> His Majesty by this means will have the reputation of giving the office to a man whose profession it belongs; Mr Hamilton would enjoy the profitts of the place without subjecting himself to the Disagreeable Dilemma, he will be in, if he holds the office in his name ... and I should have the reputation of being in his Majesty's Service, and satisfying a pride of being surveyor to those buildings erected by my ancestors.

Sadly this neat solution was not adopted, but the letter confirms that his interest was not in the salary the post would have brought.[15]

During the 1760s he formed another connection that lasted the rest of his life. Where the new bridge reached the north side at Blackfriars, its arrival meant some extra land for the riverside wharf owners because there was to be a new embankment. A painting from 1750 shows that just upstream from the mouth of the Fleet was a large wharf with a crane, some sheds, and neat stacks of tree trunks, and behind it all an old red brick house whose windows looked over the river.[16] This was the office of London's largest waterworks, the New River Company, which had supplied piped water to businesses and private houses in the City and West End since the completion of its aqueduct in 1613. *The Governor and Company of the New River brought from Chadwell and Amwell to London*, as it was formally called, was a profitable private enterprise that had its origins in the time of Queen Elizabeth. It had been given its charter by James I in 1619 and he had acquired half its shares in the process, but Charles I who inherited them had sold them, and the whole company was now in the hands of a small number of very wealthy individuals. The water it supplied was not the foul liquid of the Thames, but fresh water from Hertfordshire springs, in practice augmented by a slightly less wholesome inflow from the upper part of the River Lea just below the town of Hertford. This relatively clean water flowed almost forty miles along a winding manmade channel, almost level, that followed the contours from its source to settling reservoirs at New River Head by Sadler's Wells on Islington Hill. From there the water was distributed all over London in a network of massive pipes bored from tree trunks. Each trunk was pointed at one end like a pencil and had a tapering round socket at the other, so that each fitted into the next and made a workable if porous pipeline. They were buried along the centre of the streets, and from them small lead pipes ran to the houses of those prepared to pay a guinea a year for an intermittent supply. Canaletto drew at least one view of London from Islington Hill that showed the waterworks prominently in the foreground, so perhaps it was a version of that picture Mylne bought.[17]

The wooden pipes had a fairly short life before they rotted – anything from four years to twenty-five depending on ground conditions, Mylne later noted. This meant that the Company had a constant need for long straight sections of elm and other timber, to replace rotten sections and to extend its network to new districts as London grew. Moving tree trunks along the muddy and mostly unmade roads of eighteenth century England was slow, expensive, and impossible in winter. As a result most of the timber came by water, down the Thames as well as up, so the company had a wharf, and also its horse-driven boring augers and its offices, by Mylne's new bridge.

Mylne had to negotiate with the company over the value of the land they would gain from the embankment. Because of its complex system of open aqueducts, bridges, pipelines, reservoirs and pumps, the company had its own surveyor and engineer, the versatile Henry Mill, who in 1714 had taken out a patent for what is thought to have been the world's first typewriter.[18] By now, though still in office, Mill was well over eighty, so once again Mylne was in the right time and place with almost precisely the right qualifications – in Rome he had taken a particular interest in the ancient Roman water supply, probably encouraged by Piranesi who admired it. In 1767 he was asked to take over, nominally as Mill's assistant, and on Mill's death two years later, Mylne succeeded him as surveyor and chief engineer, a post he kept for over forty years until his son William Chadwell Mylne stepped into his place a few months before his death.

As it happened the company's offices burnt down on Christmas Eve 1769. Fortunately no lives were lost, the company's deed-chest was saved and the building was insured. The opportunity was taken to move to an adjacent plot the company held and Mylne was asked to design a new House containing a court-room, offices and apartments for the clerk, together with stables, workshops and sheds. The first stone was laid in June 1770, over a deposit of coins and memorabilia, and the building soon completed to a pleasingly simple classical design, set along the edge of a narrow plot leading down to the river so that it had some river views (see illustration 29). Like so much of Mylne's work it did not long outlast him, and by 1820 was replaced by the City gas works and in 1883 the City of London School.

This appointment probably caused a change in his career. Although the salary was only £200 a year while Mill was alive, it brought his total income to about £800 a year even before taking account of any fees for architectural or engineering commissions, which were sometimes as much again. For one who had learnt the skills of careful housekeeping in Rome, this made his earnings more than adequate for his present or likely future needs. From the financial records in his diaries we know that he had a surplus income that he was investing in long term annuities years before he was paid in full for Blackfriars, and many of those investments were never used by him but were left to his children. In the three years up to 1772 he noted that he had invested £3,100, though this included household furnishings and a loan to his brother. An annual income approaching £2,000 did not make him a rich man, but it made him very comfortably off, and although he always worked hard, he did so because he chose to, and not merely for money. Indeed, he spelt out his views on the subject in a letter to his mother in 1770: 'Money, I always despised, because I could procure it by Industry.'[19]

So it came about that within a few years of arriving in London penniless and virtually unknown he was a man of substance in charge of his own affairs. Although he had salaried posts, which carried obligations to his various employers, it is clear that he always acted as an independent professional in the sense that clients told him what they wanted, and he told them, usually politely, how it would be done. Further and most important, because his income now came from several sources he was not beholden to any one of them.

Not surprisingly in these circumstances, as an intelligent single man in 1760s London he began to attend gatherings that were more than social. This part of his life is much harder to unravel, not because he tried to conceal it but because it mostly called for no entries in the kind of diary he kept. It has to be gleaned from the scanty references that happen to survive in other people's diaries or letters and inevitably much of it will never be known. As far as formal societies were concerned he was from 1762 a member of the Royal Society of Arts, which held meetings to discuss all current aspects of art and manufacture at an exciting time when England was seeing all the changes of industrialisation but before its damaging social effects were too obvious.[20] He was also elected a Fellow of the Royal Society in 1767, and this increased the range of contact he had with other men of science. Founded by Charles I, the Royal Society had Wren as one of its first members, and architects and engineers were as welcome as those such as Sir Joseph Banks and Captain Cook who were exploring the boundaries of the natural world. As it happened, Mylne was one of those who supported Cook's application to the society, eight years after his own election and was an active member. On one occasion he was able to make St Paul's available for an experiment with Banks, Solander, Watson and others to take measurements in order to ascertain the weight of the atmosphere at different heights, from the banks of the Thames to the cross on top of the dome.[21]

More intriguing, and virtually unknown, is Mylne's association with one or more of the many clubs that flourished in London's taverns and coffee-houses, and it is one of the most teasing omissions from the diaries where it is an aspect of his life that is completely unrecorded. One held meetings at Slaughter's coffee house in St Martin's Lane.

Writing in old age, Jeremy Bentham recalled the London of his youth:

> At that time, and for a good many years afterwards, there existed a sort of philosophical club, composed at first of but a small number of members, which, at different periods of its existence went, I believe, by different names, two or more, no one of which is at present in my memory. At that time, the number of

> its members was small, but antecedently to its extinction, its members, as well as its celebrity, had received considerable increase. Sir Joseph Banks, the late Dr Solander, John Hunter the surgeon, Myln the architect, the still existing and celebrated Mr Richard Lovell Edgeworth, Dr George Fordyce the physician, Jesse Ramsden the optician, Conyers the celebrated watchmaker and writer on that subject, another Conyers, Arabic Professor somewhere in Scotland, and perhaps one or two more members[22]

Mylne left just one note about Bentham, an otherwise blank sheet in his diary intriguingly headed 'Bentham's remarks on Saxon Churches.' The same year he also noted having had to pay one guinea as an unexplained 'forfeit at the club', so perhaps his head nodded after a hard day's work as young Bentham began to expound.[23]

At least one of those named by Bentham, the author Edgeworth has also given a valuable description, in which he mentions all the names on Bentham's list except the Conyers and Bentham himself, and adds six more of interest:

> I was introduced by Mr Keir into a society of literary and scientific men, who used formerly to meet once a week at Jack's Coffee House in London, and afterwards at Young Slaughter's Coffee House. Without any formal name, this meeting continued for some years to be frequented by men of real science and of distinguished merit. John Hunter was our chairman. Sir Joseph Banks, Solander, Sir C. Blagden, Dr George Fordyce, Milne, Maskelyne, Captain Cook, Sir G Shuckburgh, Lord Mulgrave, Smeaton and Ramsden, were among our numbers. Many other gentlemen of talent belonged ... but I mention only those with whom I was individually acquainted.... the first hints of discoveries, the current observations, and the mutual collision of ideas, are of important utility. The knowledge of each member of such a society becomes in time disseminated among the whole body, and a certain esprit de corps, uncontaminated with jealousy, in some degree combines the talents of numbers to forward the views of a single person .... I am not a freemason; I cannot, therefore, speak of the initiatory trials, to which a brother is subjected; but in the society of Slaughter's Coffee House we practised every means in our power, except personal insult, to try the temper and understanding of each candidate for admission. Every prejudice, which his profession or situation in life might have led him to cherish, was attacked, exposed to argument and ridicule.

In a later passage he recalled that Sir William Hamilton was also a member.[24] Among these names, Hamilton's inclusion is confirmed by a letter Mylne wrote to him in 1773, by which time he was ambassador to Naples. Mylne was preparing to publish a work on the Antiquities of Sicily – never completed – and needed some drawings from Hamilton, but then went on to mention what he called the Club of Philosophers, as scientists were generally called.

> … investigating the formation of this Globe, gains ground daily. I think it will lead to a perfect knowledge of it in time. We have had some new Ideas about the Lightning on Volcano's from a Mr Brydone lately. I do suppose you have seen his late publication on Sicily – as well as another account lately published from the German by Mr Foster, who is gone round the South Seas. If you have not yet got them, I will send you them immediately. The Club of Philosophers desire to be particularly remembered to you. A new Member of that Body, Capt Phipps, goes a voyage with 2 Frigates this season to the North Pole, if they can penetrate so far – at least their professed intention is to explore the Seas thereabouts. A Hardy Undertaking, but he is an intrepid character. Perhaps you know him, he is Lord Mulgrave's son. An Observer from the Royal Society goes along with him …[25]

The group Mylne refers to seems to be different from the better-known Philosophers' Club, which was an offshoot of the Royal Society. The voyage to the northern regions he mentioned did go ahead, and by an odd coincidence, a young midshipman who sailed with Phipps on that journey and was later said to have fought a polar bear on the ice, was destined to cuckold the aged Hamilton and later to be buried by Mylne – he was Horatio Nelson, then aged fourteen.

Another of Mylne's interests was in his own forebears, and not just the famous ones. Soon after he was settled in London he began to compile a family history that he never completed.[26] Many relatives were what he called 'bred to the sea', some as sailors or captains of merchantmen, one a ship's surgeon. One died on the coast of Guinea, another off Africa, while others were never heard of again. Great uncle James had gone to the ill-fated Darien colony, and then privateering when it failed, ending in Barbados, while others were merchants or tradesmen.

All this while William had been developing his career in Edinburgh, and two years after their father's death, in 1765, came the news that he had at last won a prize to equal Robert's. The long-discussed work was to take place that would give Edinburgh a whole new town to expand into, an elegant planned Georgian town to be built on land to the north of the old one. All that was needed was a bridge over the valley of the North Loch to make a link between old and new, and William had been awarded the contract to build it. He had won this prize not in the kind of open competition that Robert had faced in London but according to the old traditions of Edinburgh after secret discussion within the self-perpetuating Town Council, of which he was a member.[27]

Technically the construction should have presented no problems. There were just three large arches and most of its length was solid abutment. Although it was somewhat longer than Blackfriars Bridge – 1,134ft instead of 1,000 – the only real complication was that it had to be 70ft high to bridge the deep valley. As was usual in public contracts, William had to find sureties for his performance, and one of these was Robert, already sufficiently substantial. William had four years to complete the bridge, the contract price was £10,140, and he must have set to work with the feeling that at last he was moving out from Robert's shadow.

There were curious parallels between William's bridge and Robert's. Each was the major construction project of its time and place, and each was expected to be finished in 1769. Just as Robert looked out at his bridge from his house and office at Arundel Street, so William was able to look up at his from his house in Halkerston's Wynd. The Wynd represented medieval Edinburgh, a steep narrow lane leading down to a postern gate, while the new bridge towering overhead must have seemed a symbol of change and renewal.

As work continued it all looked promising. Eventually enough had been built for pedestrians to start using it unofficially. They were warned not to, but a short cut that avoided the scramble down into the valley and up the other side was too good to resist. Perhaps the quirkiest account of what then happened on 3 August 1769 is to be found in the journal of Darcy, Lady Maxwell, a very devout resident, and one who could have taught Robert Mylne a thing or two about keeping a diary:

> The Lord, who is continually loading me with his benefits, has twice this day remarkably interfered in my behalf ... In the evening he preserved me from broken bones to which I was exposed by a fall. A few hours after, when walking home from chapel, I witnessed a most melancholy scene occasioned by the falling in of the North Bridge. I ... was within five minutes of passing over it ... When almost in a moment, the greatest noise I ever heard (except on a similar occasion when I was remarkably preserved) filled the air. It seemed as if the pillars of nature were giving way. Instantly the cry resounded 'The bridge is fallen'.[28]

It was a disaster both for William and the town. Part of the high structure had collapsed with the loss of five lives. According to one report, part of the cause was that the foundations rested on an accumulation of rubbish thrown out of houses on that side of town over many years, while another blamed the immense load of earth placed upon the arches to bring the road level high enough. Robert hurried up to Edinburgh to see what was to be done, while the Council commissioned a report from Smeaton, John Adam and another, names that must have been galling.[29] Building the New Town had already started and people needed access to it. Payments to William were stopped until matters were set right and there were demands that he should take down much of his original work to prove it was safe. For William it was to prove the effective end of his Edinburgh career because, as later became clear, he did not have the temperament to live down such a failure.

Thus, as the 1760s came to an end Mylne's delight with his own success must have been tinged with anxiety about William's future.

1 Elevation of Tusmore, Oxfordshire, one of Mylne's grandest houses. Despite the very regular front it had a more flexible interior layout, reflecting Mylne's view that houses must serve their users' needs. Completed in 1770, it was demolished in 1960. (Courtauld Institute, Conway and Witt Libraries)

2 New River Head from the south *c.*1795. The Water House at New River Head on Islington Hill was Mylne's family home in London from 1774 until his death, and his son remained there until the 1850s. Originally built in 1613, Mylne extended it for his growing family. It can be seen in the centre of the picture behind its brick wall. The smoking tower to the left is the engine house that pumped New River water to what is now Claremont Square reservoir, to serve higher districts. Workmen in the foreground are servicing some of the vast network of wooden water mains that led all over London. (London Metropolitan Archives)

Above left: 3 Robert Mylne aged twenty-four. This engraving by Vangeliste was made in 1788 from a portrait made by Mylne's fellow-student Richard Brompton in Rome in 1757. Mylne preferred the original and wrote of this engraving: '... The Head seems not to have Brains enough; The Scull ... is too scanty... The nose is too long... The under Eye Lash is very stiff. The chin projects not far enough...' (See page 198.)

Above middle: 4 Robert Mylne in middle age, by his daughter Maria. Maria died when she was twenty-two, and her father was in his fifties when she painted the miniature from which this engraving was later made. Like all his portraits it shows a pugnacious jaw.

Above right: 5 Robert Mylne in old age, by George Dance the younger. Dance the architect's hobby was to sketch faithful resemblances of distinguished men to show 'how surprisingly nature has diversified the human countenance'. The sketches were later published.

6 Robert Mylne as a supplicant. The artist, Nathaniel Dance-Holland, was George Dance's brother and first met Mylne in Italy. Here he draws him around 1766, newly appointed as Surveyor to St Paul's and beseeching Archbishop Secker for building funds. This portrait, like all the later ones, shows Mylne with a pigtail. (Tate Britain)

Above left: 7 Thomas Mylne, d.1763. Robert's father, architect and deacon of the Edinburgh stonemasons. Almost certainly encouraged by his wife, he provided the modest allowance that helped Robert and his brother William to study in Europe.

Above middle: 8 William Mylne 1734–90. Robert's brother shown holding a drawing instrument in a dark portrait that seems to confirm the introspective personality his letters suggest.

Above right: 9 John Paterson, the committee chairman and later friend who supported Mylne's winning entry for Blackfriars Bridge. Mylne commissioned this portrait from Sir Joshua Reynolds within weeks of receiving final payment for the bridge in 1776. He also paid for an engraving to be made from it. Paterson is shown holding plans to raise funds for the bridge and other improvements.

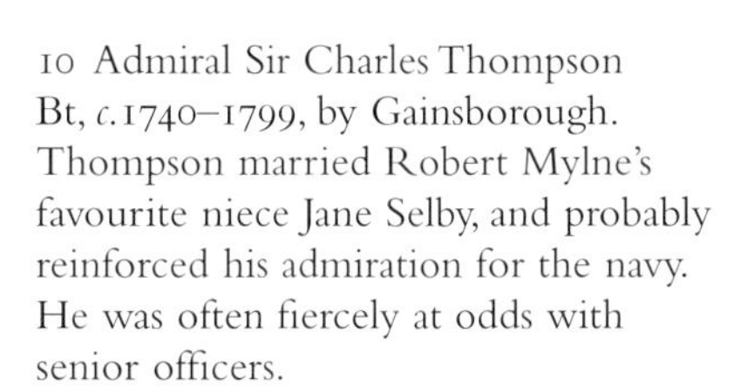

10 Admiral Sir Charles Thompson Bt, *c.*1740–1799, by Gainsborough. Thompson married Robert Mylne's favourite niece Jane Selby, and probably reinforced his admiration for the navy. He was often fiercely at odds with senior officers.

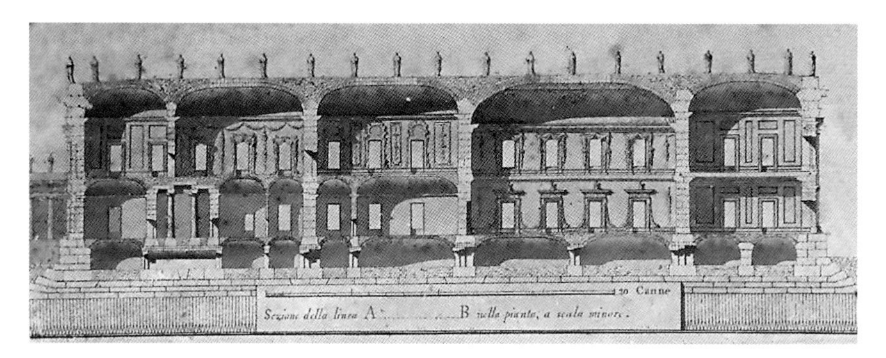

11, 12, 13 Competition design, Academy of St Luke, Rome, 1758. The competition was held every three years and the subject that year was a gallery to house the busts of famous men. Mylne won the first prize and was the first Briton ever to do so. It made his career, perhaps because he was sufficiently astute to publicise the victory in the British press. The design shows a mixture of neo-classical and French elements, and was influenced by his studies under Piranesi. (11 RIBA Library drawings collection; 12 and 13 Archivio Storico, Accademia Nazionale di San Luca, Rome)

14 Both faces of Mylne's prize medal, Rome 1758. One side shows the newly elected Pope Clement XIII, the other shows St Luke painting the Virgin Mary.

Above left: 15 John Smeaton FRS 1724–92. Smeaton and Mylne were colleagues and sometimes rivals. Mylne's design for Blackfriars Bridge was preferred to Smeaton's, and he made an adverse report when Smeaton's bridge at Hexham was swept away because, said Mylne, of inadequate foundations. In 1783 they appeared on the same side as witnesses in the Wells Harbour case, and Mylne later issued a challenge to force an apology for Smeaton as well as himself for the cross-examination they were both subjected to in that case. (See pages 138–148.)

Above right: 16 Laying the foundation stone of Blackfriars Bridge, 1760. The Lord Mayor Sir Thomas Chitty wanted the stone laid before his year of office was over, so the ceremony took place on dry land before any building began. Afterwards the press reported that Mylne had impulsively placed 'his medal from Rome' under the foundation stone. Mylne is probably the young man in wig and knee breeches facing us at the right edge of the stone. His own letters show he had actually won two medals in Rome. This may be the first example of his habit of making such deposits, which reached its climax at Nelson's funeral.

17 'Just Arriv'd from Italy': satire on Mylne's anonymous pamphlet. Mylne won the Blackfriars Bridge competition partly with the help of his anonymous pamphlet *Observations on Bridge Building*, which meticulously analysed the merits of his design and the faults of its competitors. This satire depicts him squatting rudely on the part-built bridge and discharging invective, including the title of the pamphlet and a medal labelled Rome, at his fellow competitors who are gathered below. (British Museum Trustees)

Opposite: 18 'The (Bute) interest in the (City) or the (bridge) in the (hole)', rebus satirising Lord Bute, Mylne and the bridge. In the top drawing Mylne is shown in front of the table, smartly dressed and with a rolled drawing labelled 'the best plan' under his arm. His hand is at his sword-hilt as he proclaims 'I can survey elliptically as well as draw', a reference to his controversial elliptical arches, while a sturdy man on the left warns 'Take care – I'll survey ye for ye good of ye City.' On the table a letter from Lord (Bute) reads 'He must be the man'. On the far side of the table are Paterson and others.

The rebus underneath complains of the influence of Lord Bute and Scots generally, and of the choice of Mylne to build the bridge. Most of the symbols are obvious and in 'twill (k)not be s(cott)-free', a 'knot' was the looped shoulder pad worn by porters to cushion heavy loads. Other symbols include ass, awl, bear, bee, breech, bridge, buildings, city, cott, duck (Duke), ear, eye, fool, foot, fortune, garden, hat, heart, hen, man, mouse, (no)body, one, ring, sheep's head, shoe, sword, toe, ward, well, yew. (See page 58.) (British Museum Trustees)

The Interest in the or the in the

Ye lards & ye Gents t far Nor do d
Attend my c & I'll serve ye
With lank s & broad s, both keen hell

N m nd for Y Clothes nor yet for y itch
Leave Y bannack y w th a
Away yay Ye'll - -rich

Besure y Gents b Y ped grees long
From Mc Murdo to n m nd y wrong
W w th Lord & y all m ngle among

If sm s ye have w th some nat parts
T w ll seve ye Fair sex & ye'l gain their s
For Wh ning & Cr nging are 'd scott sh parts

If Italy ere as a you've been
And ye Fa r with 1 der have seen
Make a report n ye C ly & a U'll gin

T ye Scotch d have power some s says a pity
Such a fool s ye Author of th s ish ditty
But theres m Fools he be found in ye Citty

Now 'll throw down my & I'll w th Agree
That W s noth ng nothing me
And w theres a nen t will be s Free

At each stave Song Tantarrara sing tan &c. Sold in May's Covent

Plans for a bridge at Blackfriars were put forward when the City lost trade after Westminster Bridge opened in 1750 and broke the old monopoly of London Bridge. A competition to choose a design was announced on 15 July 1759, the very day that Robert Mylne landed at Harwich after five years in Europe (see chapter 5). As can be seen in fig. 19, Mylne's design included statues of British naval heroes between the columns of each pier, an early example of his admiration for the navy. The statues were never made because of the cost. Figures 20 and 21 show some of the sixty-eight alternative designs, of which few survive.

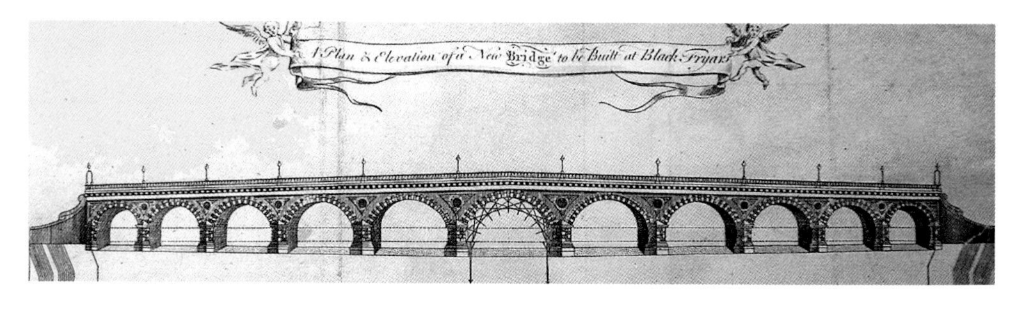

22 Piranesi's engraving of the part-built bridge. Mylne had studied under Piranesi in Rome, and sent him enough details to make this engraving. It shows the bridge at the time when the great central arch had just been finished. The Lord Mayor and sheriffs were the first to pass through it in their ceremonial barge on 1 October 1764 (see page 75). To show the scale of the work – the arch was 100 feet wide – Mylne arranged for about fifty workmen to range themselves around the arch, one to each stone, making what the *Gentleman's Magazine* called 'a very pretty effect in showing the magnificence of the arch by the comparative view of men and stones'. (Guildhall Library, City of London)

Opposite above: 19 Mylne's design for Blackfriars Bridge 1759.

Opposite middle: 20 Three designs for Blackfriars by Edward Oakley.

Opposite below: 21 A 1756 design for Blackfriars.

Three views of the caisson. The vast floating caisson, big enough to hold two layers of ten double-decker buses, was repeatedly sunk to the riverbed so that stonemasons could build the foundations of the bridge piers in dry conditions. Each time one of the underwater piers was complete the base of the caisson had to be left under it and the sides were lifted away by barges to have a new bottom fitted before the caisson was floated out to the site of the next pier. (Baldwin, 1787) (Guildhall Library, City of London)

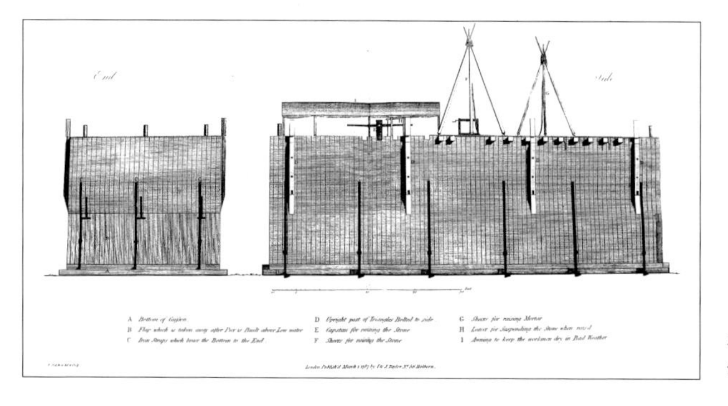

23 Side and end views of the caisson.

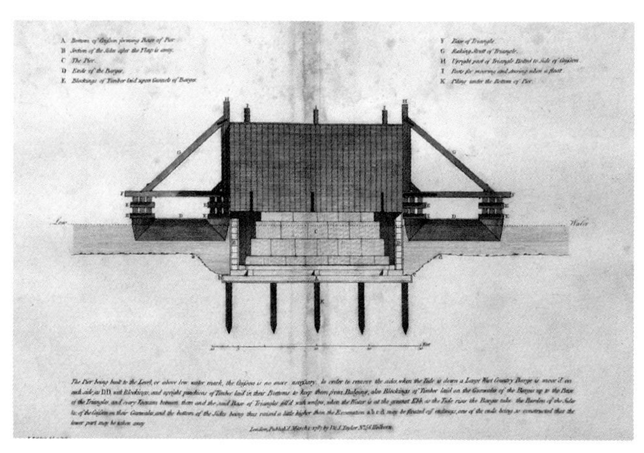

24 Raising the sides of the caisson by attaching barges to the sides at low water.

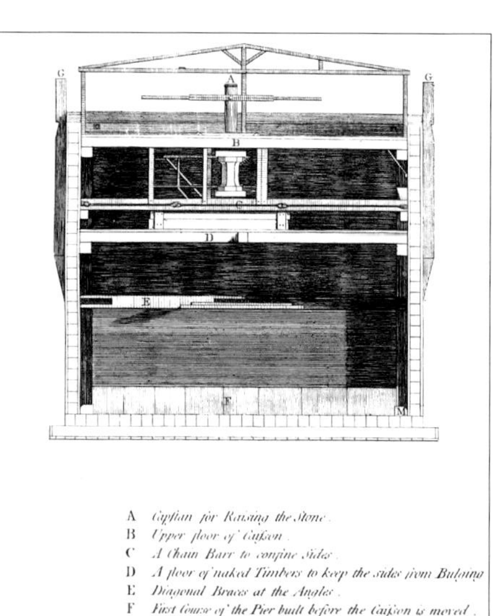

Left 25 Section through caisson.

Opposite below: 27 Centring used at Blackfriars. Mylne's design for centring to support the arches during building was 'so simple that one is surprized it has not been found out before now' and combined structural advantage with a much less wasteful use of timber than traditional methods.

Right 26 The underwater saw. Piles were driven to make firm foundations for the bridge piers. Mylne designed the saw so that two men in a barge above could cut off the tops of the piles within an inch of the riverbed once they could not be driven any further. It could be used in water 14ft deep.

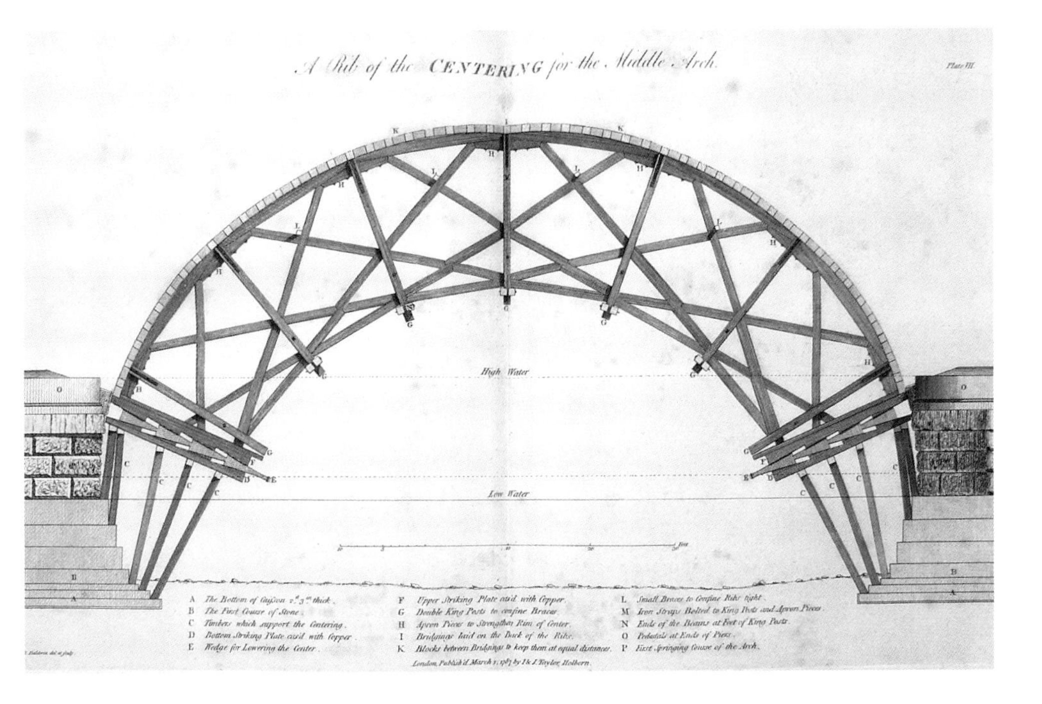

28 Mylne laid out a broad square at each end of the new bridge. This view shows Albion Place at the south end, and the curved end to the bridge parapet.

29 Elevation of new offices for New River Company, Blackfriars, 1770. The New River waterworks had its offices and timber wharf (for wooden pipes) at Blackfriars. The building burnt down on Christmas Eve 1769, and Mylne who was the company's engineer designed these new offices, to be built at right angles to the river on the site of what is now the old City of London School building. It included a court-room for the board, offices, an apartment for the clerk, stables and workshops for making wooden pipes. (Guildhall Library, City of London)

30 City of London Lying-in Hospital, 1773. Built according to the latest ideas for maternity hospitals, this building stayed in use until about 1907. It had large airy wards with wheeled beds, separate delivery rooms, and a system of natural ventilation designed to remove the foul air that was believed to spread 'impurity and infection'.

Right: 31 Elevation and floor plan for unknown house. This design shows how Mylne tried to include pleasing and unusual room shapes in his designs. This was first seen in his oval St Cecilia's concert room in Edinburgh of 1761–63, which still exists. (Courtauld Institute, Conway and Witt Libraries)

Below: 32 Floor plan for The Wick, Richmond, 1775. The Wick, which survives, has been called 'one of the most perfect of small Georgian houses' (see page 103). Mylne built it for Lady St Aubyn. The oval room at the back of each storey commands a magnificent view of the Thames from Richmond Hill. (Courtauld Institute, Conway and Witt Libraries)

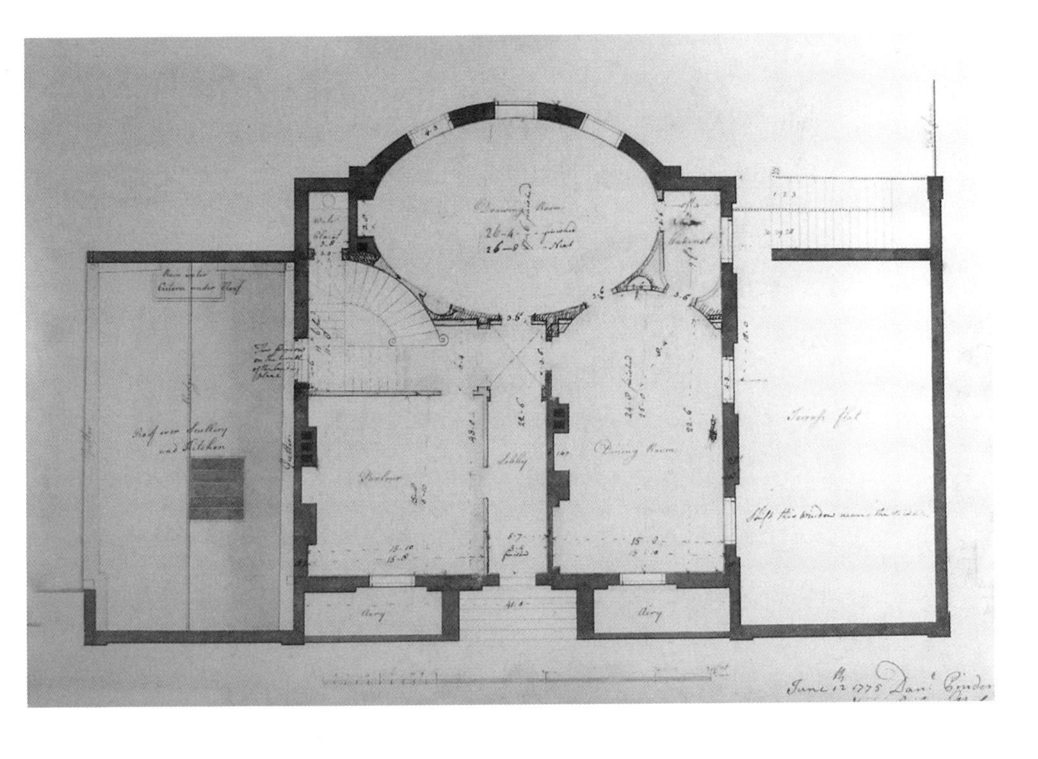

Above left: 33 Matthew Boulton: two sides of an advertising token made for the French market. The obverse is a portrait of Matthew Boulton. On the reverse an inscription in abbreviated French records Boulton's invention of a steam-driven machine to strike coins in 1788, improved in 1798, with which eight boys could strike between 400 large and 920 small coins each minute without fatigue.

Above right: 34 Matthew Boulton: medal to commemorate Battle of the Nile, 1798. This gold medal, commissioned by Nelson's prize agent, shows the fine detail possible with Boulton's coining press. It was awarded to the unrelated Bolton of HMS *Swiftsure*, hence the rougher inscription that has been added. (Victoria & Albert Museum)

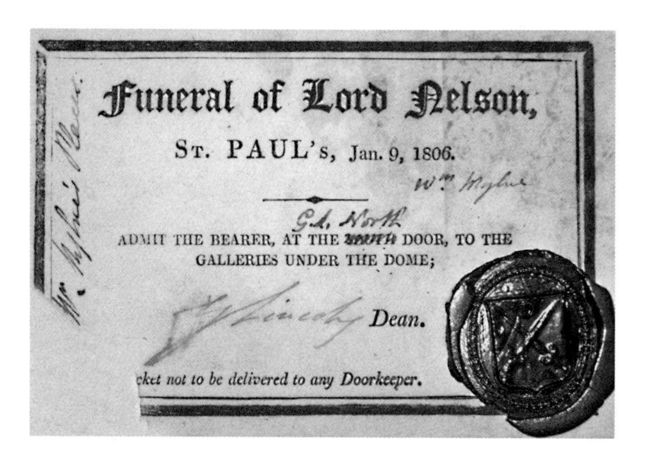

Funeral of Lord Nelson,

ST. PAUL'S, Jan. 9, 1806.

Wm Mylne

G.d North

ADMIT THE BEARER, AT THE [illegible] DOOR, TO THE GALLERIES UNDER THE DOME;

Dean.

cket not to be delivered to any Doorkeeper.

Above left: 35 Admission card to Nelson's funeral. Mylne as Surveyor of St Paul's had an allocation of tickets for the funeral, each signed by the Bishop of Lincoln who was Dean and sealed in black wax. This is his son William Mylne's ticket, and a doorkeeper has torn the corner off to prevent re-use, as far more people wanted to be present than was possible.

Above right: 36 Boulton's steam-powered coining press of 1798, still in use in the 1860s as shown in this engraving, made our standard modern coinage possible. It used the equivalent of a forty-ton hammer blow to turn a blank disc into a perfectly finished coin in a fraction of a second.

Above: 37 Nelson's funeral cortège passing Blackfriars Bridge. A fleet of boats accompanied Nelson's funeral barge from Greenwich to the Admiralty on 8 January 1806, where his coffin remained until his funeral the next day. The voyage up river was marked by a ferocious squall, but crowds still lined the banks and bridges, and can be seen on the wrong side of the parapet here at Blackfriars, viewed from the direction of St Paul's and looking towards Albion Place.

Right: 38 Nelson's tomb in its final state, *c.*1815. The base up to just below the inscription (where the artist has misspelt the abbreviation for Viscount) is Mylne's plain sarcophagus, built in 1805–06 and containing Nelson's coffin and whatever coins and medals Mylne secretly deposited beneath it. The upper part and large inscription were added about 1810 and are purely decorative. The crypt was so dark that a lantern was needed even at noon.

Left: 39 Memorial to Sir Hugh Myddelton, Great Amwell, Hertfordshire. On a small island in the New River overlooked by his new house Mylne built a memorial to Sir Hugh Myddelton, the London goldsmith who took over the unfinished New River scheme and completed it in 1613. This was one of a series of memorials of different kinds that Mylne carried out from 1800 until his death.

Below: 40 Stationers' Hall. Mylne was retained by the Stationers' Company to manage their estate and in 1800 rebuilt the east front of their hall near St Paul's to this pleasing design, which can still be seen.

8

'... this best step of my life...'

By 1770 Blackfriars Bridge had been finished to a high standard and for almost precisely the budgeted cost, and the man who designed and built it had established his own credentials in the process. Mylne could relax a little, and the decade that followed saw him branching out. By this time he was thirty-seven, and it is no surprise that marriage came next.

The wedding took place at St-Martin-in-the-fields in September 1770, and by the 10th when he wrote to his mother at Powderhall and his brother in Halkerston's Wynd, he was with his bride at Tusmore in Oxfordshire where he had designed a country house for William Fermor (see illustration 1). The letters provide a rare expression of Mylne's own feelings and beliefs, and the one to his mother is the fuller:

> My Dear Madam,
> I have within these few days entered into so new a situation of life as demands my acquainting you with it ... I am Married, and that to a Woman, who is the Wife of my choice. – In this Step, as in every other, I have had more a regard to a prospect of domestick happiness, than any other consideration. – Money, I always despised, because I could procure it by Industry. She has no fortune, nor is of any family, nor of great Interest, nor connections. The figure of her person and what degree of beauty she may be possessed of I shall leave to others to mention to you:- anything a lover says on these points are not to be regarded. All that I shall mention, is, that I think she is possessed of those qualities, which render life smooth and easy – she is industrious, active, knowing in the affairs of a family, and most business wherein a Domestick woman shines. She is good natured, easy, full of life & spirit, has some share of wit, and great readiness of conception and expression. With these qualities, which suit my soul, I love her, and was thoroughly convinced a long while agoe that she loved me. Her name is Mary Home, the Daughter of a Surgeon to a Regiment of light Horse... As I know you take a great part in my happiness, I think it equally my Duty, as well as it is my wishes, to acquaint you with this best step of my life...[1]

News of the marriage was not as welcome as might be expected. Mylne's younger sister Anne, still unmarried and living crossly with her mother in the shabby isolation of Powderhall, wrote to a woman friend:

> My brother's marriage is not such a joyous event as you imagine, he has unfortunately joined with a family that is the aversion of Mama and all our friends, the young Lady is pretty though not a Beauty – a great share of vivacity and sharp judgment, her father is a surgeon to a regiment had 7 children and only his pay to support them – So that neither theire education nor Appearance, though Both were greatly above their Circumstances could be striking – I cannot tell you what surprise and uneasiness this affair has Cost us all – my Brother naturally haughty made all the world amaze, as he must undoubtedly support the whole family, through time ... a family whose pride, poverty and Show renderd them Contemptible to all that know them – you will think me severe – But they behavd very badly to me, By endeavouring to make a Coolness Between us, and they to well suceeded.[2]

The couple's first home was at Arundel Street, and unfortunately there is no description of domestic life there apart from what can be gleaned by the very limited diary jottings. Within three weeks of the marriage Mylne noted that they had 'Old Mill, Mr Cole and various people to Dine &c in the Course of the Week,' but there are no other entries like that. 'Old Mill' was Henry Mill, eighty-seven by then but still nominally engineer to the New River Company. He died that Christmas, so that Mylne became his successor as surveyor and chief engineer, and was later given the use of the Water House at New River Head, a mile to the north of Blackfriars.

Before that position was confirmed the couple had a house near Croydon, from April 1771, about which nothing is known apart from brief diary notes of the cost of furnishings, gardening and housekeeping there. It may have been a country retreat where they only spent part of the time, or it may be that once there were children in the offing it seemed best to keep them away from the smoke, dirt and disease of London.

A French visitor of that time who described London with the insight of an outsider thought it a city where, even in the new districts, the streets were always filthy despite the constant use of carts to carry away the dirt, the result of the level ground absorbing the rain and turning to mud. He said there was smoke and constant fog for the eight months of winter, caused by the coal burnt in homes and workshops. Makers of glass and pottery, blacksmiths and gunsmiths, were all attracted to London by the availability of sea coal and the proximity of shops to sell their products, which saved the cost of baling them up and transporting them. 'If the increase of London proceeds as far as it may,' he wrote, 'the inhabitants must at last bid adieu to all hopes of ever seeing the sun. The smoke, being loaded with terrestrial particles, and rolling in a thick, heavy atmosphere, forms a cloud, which envelops London like a mantle.' The smoke also had a ruinous effect on clothes, books and even stonework – the masonry of old Somerset House resembled filigree-work.[3]

Perhaps the only description of the Mylnes as a couple at that time is in a letter from a writer called John Gray to Tobias Smollett, who was dying of consumption in Italy. Gray had met Mylne at a friend's house and noted that he had a handsome young wife but was grown very fat, that he kept 'an elegant chariot' but matched their mutual friend Dr Armstrong for arrogance and was said to ask a good price for his opinion.[4]

As Gray suggested, Mary was younger than Mylne, being twenty-two to his thirty-seven when they married. Their first child Maria was born in 1772, by which time Croydon was their second home. In the first five years of his marriage the diary entries average about one every five days, and these are often in clusters leaving whole months blank elsewhere. In December 1773 he noted in his diary, 'Family moved from Croydon', and the following 30 May, 'Removed family to Islington'. He was based in London for his work and roads in winter became impassably muddy, so this may have been a final move from Croydon to Arundel Street for the winter season followed by the start of life at New River Head. As he later wrote to a friend, he had left Arundel Street and the follies of London for the space and quietness of New River Head.[5] Around that time he sublet the various parts of the Arundel Street premises. The Honourable John St John, a barrister MP, took the best rooms overlooking the river for £90 a year, while other parts went to other tenants.

Meanwhile the family was growing. Little Maria was followed by three more daughters in successive years, Emilia, Harriet and Caroline. Then their mother had a miscarriage, followed by another in August 1777, when she lost a boy and Robert wrote to his brother, 'this life is chequered, but we must make the most on't. We have no choice and it is useless to repine. It happened a Month agoe. She is quite Stout again.'[6]

Two years later in December 1779 the first son was born, and baptised as Robert.

As well as his new family, a new group became part of Mylne's routine, and remained so for the rest of his life. At that time England had no organisation for the fast-growing profession of engineers. Some, like Smeaton, were virtually self-taught, while others had evolved from a variety of more specialised trades. Mylne was one of those who thought this was unsatisfactory, and in 1771 he and Smeaton were among the founding members of the Society of Civil Engineers. Reminiscing about it years later, Mylne wrote how Britain in about 1760 found itself in a new era for the arts and sciences, which all began to move forward rapidly and obviously, while developments in science helped manufacturing to expand. The need to site industries in new locations convenient for raw materials and good sources of labour, instead of being 'plagued with the miserable little politics of corporate towns, and the wages of their extravagant workmen' had led to the building of canals, resulting in 'wonderful works, not of pompous or useless magnificence, but of real utility'. Meanwhile the country's ancient and neglected harbours and seaports had become unable to cope with the increase in naval power and foreign commerce until they were improved.

These things, he wrote, had given rise to a new profession and order of men, called civil engineers, long recognised as a profession in all the polished nations of Europe, who had academies for such learning while Britain had nothing beyond ordinary schools to teach it. Those who already practiced the profession sometimes met when they attended the courts to give evidence on opposing sides, but knew little of each other. One gentleman had then suggested to Smeaton – and it may well have been Mylne himself, although he does not spell it out – that it would be better to have occasional meetings, so that 'the sharp edges of their minds might be rubbed off' by a closer communication of ideas, which might promote the true ends of public works without the jostling and rudeness that he said was too common where some unworthy lawyers became involved, and where witnesses might be pushed too far in the interests of one side or the other.[7]

This led to the forming of the Society of Civil Engineers. At first they met in a tavern in Holborn on Friday evenings and later at other inns including Mylne's York Tavern and Coffee House, to discuss their mutual interests after the week's labours. The meetings were sometimes so poorly attended that Mylne or one of the others found himself the only one present, but the society was still a major step in the development of the modern professional institution. After Smeaton's death it was reorganised and Mylne was its treasurer until his death. As well as working engineers it had honorary members including men of science like Banks and Priestley. Mylne was thus one of those who founded the profession, and it is another example of his willingness to serve the greater good.

In August 1772 Mylne took leave from the New River and set off for Scotland as he did most years. This time Mary was with him and they travelled in their post-chaise. The journey north was taken more slowly than usual, and included stops at Bishop Auckland, Durham and Newcastle at each of which Mylne had work in progress. It took eleven days to reach Edinburgh where they stayed a week, which may have given Mary a chance to meet the wider family and Robert a chance to assess his brother's continuing problems with the North Bridge. They then went north to visit Blair Atholl, the scene of Robert's work as a carver before he went to Europe. From there they set off west across Scotland on the new military roads, and after three days arrived at Inveraray, a small town on the north side of Loch Fyne, to stay at the Duke of Argyll's seat at Inveraray Castle. The Duke had just succeeded his father and he and his wife were keen to have the interior remodelled, so Mylne who had done work for him in London spent a day surveying the castle and apartments from top to bottom.

The visit brought one of those odd chances that recur in Mylne's career. That night a storm of great ferocity tore down trees that had stood for a hundred years, swept away bridges and left the military roads impassable.[8] In the morning there were six local bridges down as well as all the cascades and minor bridges around the castle, so Mylne was able to dispense some advice and consolidate his standing further. It was to be the first of many such visits over the next thirty years. The Duke was keen both to enrich the castle interior and to continue improvements to the little town just outside his gates. Inveraray is tiny and remote, isolated by the arms of sea lochs that cut far inland, but it was the administrative centre for the region and had regular visits from judges on circuit and their followers. Earlier Dukes had started an improvement scheme with help from the Adam family, but Mylne was now the preferred architect, and in the years that followed he organised a vast range of improvements to the castle, its farms and the town.

The Mylnes then returned south by way of Dumbarton, where the town made them both Burgesses, and the whole trip must have been a memorable one. On such travels Mylne sometimes noted the small purchases of trinkets for his children or nieces and household supplies such as Stilton or Gloucester cheeses, mustard or barrels of herring, and one has the impression that he quite liked rattling round the country in his smart chaise.[9]

All this time in Edinburgh, William struggled to complete the North Bridge after its partial collapse, while Robert did his best from London to sort out its finances, in which he was embroiled as surety. William's problems were starting to overwhelm him. Public anxiety about the structure extended even to the parts that had not collapsed, and he had to oversee some unnecessary and costly demolition and rebuilding. We know from letters that he continued to suffer from poor health and sometimes took to his bed for lengthy periods. He also had a sensitive nature, and in such a small community must often have thought he saw derision in people's faces when he walked the streets. Not only was his home in Halkerston's Wynd right below the ruined bridge, but it happened to be the first part of the inconvenient alternative route that people had to go down to reach the valley bottom and then the upward slope that led to the new town. On frosty nights its slippery steepness was bad enough for pedestrians but even worse for the chairbearers with their sedans, and a thick coating of gritty cinders was often needed to make it passable. William must have had many glowering looks.[10] His sense of failure must have been worsened by the obvious contrast with Robert's triumph at Blackfriars, and all the more so because Robert had backed him financially and was now sharing the consequences. Eventually in May 1773 he broke and fled, taking only his dog Mungo with him, and leaving a letter to tell his mother she was welcome to the furniture and silver he had left at Halkerston's Wynd. For three months he seemed to have disappeared, and then turned up in London. His sister Anne was in on his plans and took care of young Willy, a son he had accidentally fathered on a local girl, but he could not face the advice Robert might have given him and was careful to keep out of his way. His plan was to emigrate to the American colonies as a planter or an architect, and he had a few possessions sent down to London by sea while he eked out his slender funds waiting for a suitable ship.

Parts of his travels over the next three years are vividly told in the long letters he had plenty of time to write to Anne living with their ageing mother at Powderhall. In London he was constantly aware of Robert's presence, or *His Honour*, as he and Anne sarcastically called their elder brother. When he went to the quayside to collect his trunk, box and gun from the Leith coaster one of the crew said there was also a cask for Mr Mylne – but it turned out to be wine coincidentally sent down for Robert, who sometimes ordered from Edinburgh merchants. The same day, emerging from his lodgings he had to step back sharply as he saw – or perhaps imagined he saw – Robert walking down the street towards him. He spent much of his time out walking, to avoid spending money, and one day his steps took him to Islington, where, as he told Anne, 'thought I, if his honour is at the Water works and we meet, there will be a fine piece of work.' One day he crossed Blackfriars Bridge, and told Anne that 'to give the devil his due' it was a noble work, that nothing could be better done and that it would far eclipse Westminster when the avenues on either side were finished.[11]

Instead of seeing Robert he wrote him a letter, now missing, and left it to be posted once his ship had set sail, along with a power of attorney to deal with his affairs. Once he reached America he and Robert corresponded, and Robert wrote that he could have found him a job. He said William had been wrong to think he would have been against his going abroad, but he would have recommended Italy, and had put a notice in the papers to say William had gone back there to resume his studies. William told Anne he had been very bad at the time he received that – he suffered from a recurring illness – and had written back telling Robert to look upon him 'as a man that was dead to him and to my country.'[12] Despite this gloomy note, the American trip was not a success, and within two years William was back in London. There he re-established his friendship with Robert, who was able to find him a job running the Dublin Waterworks that he held successfully for many years.[13]

The year after William's flight, Mylne was on one of his annual visits to Edinburgh where Boswell was now settled as an advocate, and Boswell records two meetings that Mylne predictably fails to note. The first was when they both dined at Alexander Donaldson's, the bookseller and publisher. Mylne had news of Dr Johnson in London, saying that he had taken up drink again and favoured wine. Another guest, a naval captain, said he had drunk no fermented liquor for years past and was much the better for it, having more stamina and needing four hours less sleep. Mylne said that he too had tried abstinence, but it didn't suit his constitution. Donaldson's claret was very good, noted Boswell, who 'loved the liquor, sucked it, and found it salutary'. They talked about capital punishment

– Boswell had a client awaiting the hangman and was preoccupied with the subject – as well as the rebellious colonists of Boston and their taxes, but he doesn't tell us Mylne's views. Two days later it was Donaldson and Mylne who dined with others at Boswell's. Unfortunately Boswell was worried that his daughter might catch smallpox from a servant whose child had it. As a result he tells us nothing of that day's conversation, merely that while the others took tea, Boswell drank too much claret and ended the evening being 'monstrously passionate' with his wife.[14]

While William was still away, Anne, who was about thirty, had found a man. Her disposition was never the sunniest, and she wrote to tell William how necessity had driven her to it. She had realised 'with horror and astonishment the small pittance that you left was all that poor Willy and I had to trust to', while she 'shuddered at the bare idea of being in any way dependent' on her older sister, and their mother 'in all her illness seldom expressed any anxiety for my future subsistence' until she persuaded her to sign a deed transferring property to Anne. She had managed to fall out with Robert over his marriage, and had later refused a payment he sent. Now she wrote to tell William how she had accepted a proposal from Sir John Gordon, who she said was an honest fellow near to fifty, a good soldier-like figure, learned, well bred and sober. He had a small fortune and a retired Captain's half pay, and would settle £100 a year on her if she survived him. The necessity of accepting had, she said, driven all before it, and she had even resolved to make him as good a wife as she could.[15]

As for Robert, his practice continued to grow. He still had good contacts in Rome, and regularly imported Italian *pozzolana*, a volcanic ash used to make a cement that would set under water, as well as arranging imports of pictures, wine and marble for himself and friends. In London he built another major work, the City of London Lying-in Hospital, now demolished, and also designed a hospital for Belfast, which still stands. As well as St Paul's he surveyed the cathedrals at Canterbury, Rochester, Durham and St Alban's, where there is a modern carving of his head. He designed many houses, some in Shropshire and one for a client in Powys, where he noted 'no expenses or charges to Mr Owen, as he made me a present of a pair of horses.' His private practice continued to grow, with the Bishops of Bristol, Lichfield and Durham added to his list of churchmen. At Newcastle he designed a new bridge, with others at Tonbridge, Yarmouth and elsewhere.

At Richmond he designed The Wick, a house for Lady St Aubyn that has survived almost unchanged (see illustration 32). In a superb hillside position overlooking the Thames, its most striking feature is the large oval room facing the river on each of the three storeys, which provides a dining room at ground level, a drawing room above it and the best bedroom above that. Christopher Hussey thought it one of the most perfect of small Georgian houses.[16]

Near Croydon he built Addington Lodge, a stylish country house for Barlow Trecothick, who had been Lord Mayor when Blackfriars Bridge was finished, and years later after Trecothick's death refurbished it for the Archbishop of Canterbury. It is clear he managed to combine the careers of architect and engineer, and even that of designer, as when he provided Matthew Boulton with designs for silverware. He was also becoming known as an expert witness, and gave evidence for James Watt before the House of Commons in support of his application for a patent for his steam engine – the major invention of the century.[17] In Scotland he advised on the use of engines for draining lead-mines.

The City of London Lying-in Hospital, on an irregular plot at the junction of Old Street and City Road leased from St Bartholomew's Hospital, was designed according to the latest medical thinking recommended by Dr Hulme. It was to have:

> ... long spacious rooms with a Range of Beds moving on Castors ... a Fireplace in the middle ... Windows at each end with Sashes letting down at the top ... a Ventilator fixed in each Bay. By these means the Rooms would be easily kept clean and pure ... there will be a thorough and perpetual Ventilation of Air from each end of the room to the middle thereof which like a Broom will sweep before it all Impurity and Infection, and there meeting with the rarified Air about the Fire-place It will form a little Whirlwind or vortex of foul Air, which will reach into the Chimney and be carried away with the Smoak ... [also] some small Wards ... for the delivery of all the women ... as soon as ever the Woman is delivered she would have everything neat and clean put about her and then be wheeled into the Great Ward ... and be put into a clean warm bed where she will be greatly refreshed and comforted by having everything clean fresh and dry about her. It may be said the good women do not mind much the Noise of their Fellow Sufferers in Labour as they know it is in general soon over and as soon forgotten ... sometimes there is more Merriment at a Labour than at a Feast ... but he thought it much better to have the Labour beds in a detached Ward ... also ... for ... patients seized with Fits or Madness ... or in difficult Labours where the Woman is obliged to be delivered by Instruments ... or attacked with a raging Fever or the like.

Mylne's designs were ready in August 1770, and in October the foundation stone was laid over a deposit of coins, after which the Lord Mayor gave a dinner in the 'most elegant and polite manner' as well as a donation to the hospital. Mylne's finished building had a pleasingly elaborate elevation, and the hospital opened in 1773 and remained in use until 1907 (see illustration 30).[18]

A puzzling setback around this time was the difficulty Mylne had in obtaining his final payment for Blackfriars Bridge. The bridge committee, which formed a large and influential part of the City's Common Council, had always intended that his salary as surveyor would be subject to a final bonus on satisfactory completion to make his fees up to the normal rate for such public works. It was well established that this was about five per cent of the cost of the wages and one per cent of the cost of materials, and the City had paid this to its own surveyor on other major projects such as Newgate Prison, the Mansion House, and improvements to London Bridge.

Mylne accordingly applied to the committee for the extra payment in December 1770, by which time the bridge had been open to all classes of traffic for more than a year, was highly regarded, and by contrast with the far more costly Westminster Bridge had shown no structural problems. The committee needed the approval of the Court of Common Council before making such a payment, and forwarded the application with their unanimous support. They pointed out that the payment had always been intended and that Mylne had performed the different roles of architect, engineer and surveyor for the bridge as well as most of the necessary work for the approach roads, embankment and temporary bridge, including the covering of the River Fleet. Despite this the necessary majority was not forthcoming from the Court, and the request was refused in June 1771. The reason is nowhere disclosed in the minutes, and it may simply be that some members felt that the sizeable salary should suffice and that they need not pay more unless they were legally obliged to. Mylne was not one to give up, and renewed his petition, again supported by the committee, until eventually the City took advice from its lawyers, including the Recorder of London and Common Serjeant. They considered it significant that when the City had applied to Parliament for the power to raise further funds for the purpose of completing the approach roads, embankments and some property purchases, the sums due to Mylne based on the five per cent payment had been among the costs listed. Because of this they advised that because Parliament had approved the application Mylne was entitled to his five per cent at least for those works. This was sufficient to decide the whole matter, and in 1776 Mylne received the sum of £4,209 that he had applied for more than five years earlier, an amount that he noted in his diary like any other fee.[19]

Although Mylne continued to be surveyor to the bridge for the rest of his life, he was obviously pleased to have its construction brought to this satisfactory conclusion, as there had always been a gentleman's agreement that the fee would be paid. He had long been solvent by this time and had no particular need of the money, which he invested in property and shares, but he marked the occasion in other ways. From Sir Joshua Reynolds he commissioned a portrait of his friend John Paterson the committee's chairman, 'whom I esteem and to whom I owe so much' as he had once written of him (see illustration 9).[20] It shows him holding the proposals for funding the bridge, and Mylne had engravings of it made for wider distribution, while he also gave a large dinner and an entertainment for the bridge committee. He took the opportunity to cancel two thirds of the debt his brother owed him for bills he had had to pay as a surety for the North Bridge, leaving a debt of just £265. This was more generous than it may sound today – the £500 he remitted would have sufficed to commission fourteen more portraits by Sir Joshua Reynolds.[21]

Throughout this period Mylne was still surveyor of the fabric and paymaster of the works at St Paul's. A French visitor complained of the strictness with which it was kept shut, except at the times of service. 'This custom of keeping out all external air occasions a dampness, which renders the church very unhealthy: the fogs that fill it during the whole winter, continue sometimes in that inclosure to the end of the month of May. The only reason for its being shut so strictly proceeds from a sordid interest… Except at the hours of service, there is no admittance without paying threepence; a contribution levied equally on natives and foreigners… the money is for the benefit of the lower choir of St Paul's.'[22]

Although the general running of the cathedral was in the hands of the Dean and chapter, the money for maintaining the fabric was provided separately by Parliament, and kept in a trust fund controlled by the Archbishop of Canterbury, the Bishop of London and the Lord Mayor, none of whom was otherwise involved with the running of the Cathedral. One of Mylne's duties was to check workmen's bills, which could not be paid until he had approved them. Although many of the records from that time have been lost the story of one characteristic skirmish he had with the Dean and Chapter survives among old records at Lambeth Palace and is very revealing.

It began in 1776, by which time he had held his appointment to the cathedral for ten years. The Dean and Chapter had appointed a painter called William Powell to do any necessary painting, but Mylne noticed that his charges were higher than other reputable tradesmen. He showed Powell lower estimates from other painters for the same work, but Powell refused to lower his prices, while the Dean insisted he should be used. Mylne felt there was some impropriety in approving the payments but the Dean gave him no option. Mylne accordingly wrote to the Bishop of London, setting out these facts and asking for his guidance. How should he as their surveyor act? He explained that he was about to set off to Newcastle, where he was building a bridge for the Bishop of Durham. He would be away for some weeks, and on his return proposed to call on the trustees to learn their views on the matter. In the meantime, he explained, he had accepted one estimate from Powell, where his tender had been the lowest. This was to gild and repaint the two clock-faces in the south tower.

The Bishop's reply to Mylne has not survived, but Powell's bill has, and is in a form clearly calculated to escalate the dispute. First it sets out the agreed price for painting the two dial plates, but then sets out an additional item, for 'extra work' painting the astragal – the rim around the dial on which the figures showing the minutes appeared. According to Powell his estimate had only been for painting the face of the dial. Mylne would not pay the extra charge, saying that it was quite obviously included in the main work, and further pointed out that even if it had been a legitimate charge – which it was not – Powell's bill wildly overstated the area of the astragal, and the extra charge was more than doubled as a result. This he had checked in typical Mylne fashion by having himself hauled up the tower to measure it. It seems clear that the Dean and Chapter felt the time had come to stamp their authority on their surveyor, and he was told to pay the bill. When he produced that year's accounts, he did not include Powell's extra charge, and no one but he was entitled to add it. Tempers became frayed but neither side would give way. There was then a hearing before the Dean and Chapter, when Powell gave his version of events and was supported by another tradesman. Mylne was then asked to pay Powell in the following year's accounts.

The matter dragged on and in February 1778, Mylne wrote a long letter to the Dean and Chapter setting out his position. It would be improper for him to admit Powell's charge into the current year's accounts when it was disallowed in the previous year and nothing had changed. Before this could happen, he would have to swear to its fairness and this he could not do. That would be to pretend there had been some kind of mistake, when there had been none. The astragal was as much part of the dial as any other part of it. There had been no mistake on Powell's part. It had been a trick, he wrote, of the kind that forward tradesmen played upon unwary gentlemen every day, who under pretence of the agreement vamp out their bill with every extra article that could be mustered by the use of tortuous and twisted meanings. In fact, said Mylne, there were other objections that he could have made to the bill, but he had chosen to offer settlement on a reasonable footing. By the terms of the commission that had appointed him, he had no right to recommend any payment beyond this. Moreover the Dean had required Mylne to agree to certain works without seeking his opinion on them. Powell had thus received encouragement, which had made him act in the way he did and to go further than he should have done. Workmen had to accept some ruling on prices. If not, workmen who were otherwise perfectly good would make immoderate demands. All reasonable workmen accepted this and submitted to the surveyor's view if there was a dispute – indeed at another public building in his care, Greenwich Hospital, the agreement gave them no option. Mylne set out a list of how Powell's demands had varied through time. He could say from personal experience that at Greenwich the same work was done for twenty per cent less. It had been said that he wanted to employ his own chosen tradesmen. This was not so, he had never recommended anyone, nor been consulted on the subject, though he had been much plagued by some who had worked for the Cathedral before his time and seemed to believe they had some kind of automatic right to continue to do so. There were some who were improperly employed and whose behaviour merited dismissal. Mylne regretted that his advice was not sought on such things. The previous hearing before the Chapter had not been conducted fairly and he had not been given a proper opportunity then to answer the allegations made. That was why he now set out his position in writing. Far from deserving the strictures made, since his appointment he had managed to increase the Cathedral's modest building fund so that it would be better able to pay for works of reconstruction that were becoming necessary.[23]

Months passed, and on 12 November 1778 the Dean brought the Archbishop of Canterbury into the matter, writing a long and angry letter. Mylne had given him and the Chapter trouble and opposition, and was now persecuting and oppressing an honest tradesman of universal good character. He was trying to 'defraud Mr Powell' of the work that was properly his. They had written Mylne a note asking him why he had appointed another painter, ending with the words 'an answer is expected' but answer came there none. The previous year he had refused to bring Powell's bill into the accounts. This year they had suggested that he could properly put it through by using the formula 'Omitted in the last year's account, and brought into this year at the request of the Dean and Chapter,' but this he refused to do. He had sent them 'a long written answer' giving his reasons, but, said the Dean, it contained 'two falsitys as facts,' though he did not say what these were. He had referred to the Dean in 'such unbecoming language' that the Chapter reproved him for it. All over a few paltry pounds, and despite 'the ease and convenience, and at the request of the Dean and Chapter'.

Matters had now gone from bad to worse, the letter explained. Mr Powell's only way to 'recover his rights and his character' was to sue Mylne, and he had begun such an action. Indeed he might have sued the Dean and Chapter with certainty of success, but out of regard for them had gone after Mylne. If the case was heard, not only would the names of Dean and Chapter be bandied about in public, but they might even be called as witnesses! To try and avoid this they had again approached Mylne with the suggestion that he should pass the bill 'at their request', in which case Powell would drop his action. The Bishop of London had been consulted and thought this was a very fair and reasonable solution. One of the Chapter passed the offer on to Mylne, but he would only say that he wanted the matter aired in court, and would settle only if Powell was to admit he had been in the wrong. If so, he would be prepared to consider employing him again at St Paul's. Even when he was told the Bishop of London approved the compromise, 'this seemed to make little impression on him'.

Now, continued the Dean, the deputy surveyor and – perhaps surprisingly to a modern reader – Powell the painter, had gone together to see Mylne at his home, apparently to convey the Archbishop's views that the bill must be paid. To their dismay he did not invite them in, spoke to them only on the doorstep, and that in such shocking terms that they immediately had to retire to a coffee house to make a note of the conversation while it was still fresh in their minds – a note said to be enclosed with the letter, but now sadly missing. Now the Bishop of Lincoln was saying that for his part he was not going to end the matter by paying Powell's bill, and he hoped they would turn Mylne out. The letter ended with the writer's views in the matter. He could not see how it was possible for Dean and Chapter to cooperate with this man any longer. Now that he had seen how Mylne slighted even the Archbishop's authority, his rudeness and insolence to the rest of them seems less surprising. For his own part, he would not refuse to examine the current year's accounts, but apart from that, in view of all the indignities and insults he had received, would ask to be excused from ever having any transaction, or holding any communication, with Mylne if he continued to hold the office of which he was clearly so unworthy.

It seems that it was not until January 1779 that Mylne was shown the details of the allegation that he had uttered two falsities – one to do with the measuring of the astragal, and one as to whether another painter had admitted privately to him that the astragal should have been included in painting the dial. He immediately wrote out what he described as a solemn declaration, that he said he would support in an affidavit. It set out in great detail why nothing he had said was a falsehood, and named witnesses who could prove the facts. Further he had asked James Adam, the King's architect, to give his opinion, which he would have given in evidence if the matter had ever gone to trial, and he said that it was a mere quibble to say the astragal was not included, that the figure in his original estimate was ample payment for all the work, and that Adam and indeed any surveyor of repute would have made the same refusal as Mylne had done. Mylne was sorry to observe, he said, how much the worthy Dean had been misinformed by the artful colouring of an interested person – the more so when it threw a wholly undeserved imputation on the character of one who had greatly improved the cathedral and safeguarded its funds.[24]

The correspondence quoted is all among the papers of Frederick Cornwallis, the recently appointed Archbishop of Canterbury. He had formerly been Dean of St Paul's at the time of Mylne's appointment and perhaps knew him all too well. There is one further document on the topic, and it remained among the Mylne family papers until 1917, when it was among a bundle relating to St Paul's presented to Lambeth Palace by his great-grandson. It is a small slip in the form of an order signed by the Archbishop of Canterbury and the Bishop of London, and it is clear from its date that it must be the message that the deputy surveyor and Powell the painter had scuttled to give Mylne at his home, before their hasty note-making retreat to the coffee house:

> Ordered: That Mr Mylne, the Surveyor, do pay the whole of Mr Powell the Painter's Bill, and bring it into this Year's Account before the Account is signed by the Chapter and audited.
> November 4th 1778
> Fred: Cant
> R. London[25]

What it shows is how resolute Mylne could be on a point of principle. The terms of his appointment spelled out that he had to authorise payment in such a case if two of the trustees ordered him to do so, but until then he was immovable and had made his point.

Around this time there was one sad piece of family news, and in April 1778 Robert wrote to pass it on to William in Dublin. He had just heard that their mother had paid what he called 'her last debt to nature'. His diary as we know was never very full, but there had already been regular entries that year. After the one recording her death there were none for over three months, and that may be a measure of his grief.

All this time Mylne continued to give regular attention to the running of the New River water system. One of the many problems in supplying London with New River water was inherent in the low elevation of its water source. Other cities often have a source of water available in nearby hills to simplify the engineer's task, but the flat contours of the counties round London meant that the springs at Amwell and Chadwell were low-lying. Accordingly the forty-mile canal, completed in 1613, that diverted them to London was designed to follow the contours so carefully that it fell by an average of only 5in a mile. Despite this, by the time the water reached London its reservoir was barely halfway up Islington Hill. This was high enough to run down to the older parts of the City and west end, but many of the newer districts on higher ground around Oxford Street could not be supplied with more than a trickle. As a result the talented water engineer George Sorocold had devised a new system in 1708, digging a large new reservoir on the very top of the hill, a few hundred yards to the north. That was high enough to serve all the new districts. The problem was that all its water had to be pumped up from the original Round Pond, and pumping technology was still primitive. Sorocold's solution was a six-sailed windmill at New River Head working pumps that forced water to the Upper Pond. For times of calm he also provided a horse-engine in the base of the mill tower, theoretically a very neat solution. When the wind blew the horses could graze the surrounding fields, eating less and thereby costing less while the wind did their work. When it dropped, the refreshed horses could be harnessed to the pumps until a new breeze sprang up. Sadly the mill was one of Sorocold's rare failures. The windmill itself was inefficient, and after the sails were twice blown off in sudden storms the company decided that the cost of renewing them again was not justified. Worse, the base of the brick tower was too small for an efficient horse mill and could not be enlarged. As the horses worked they were pushing at an angle instead of straight ahead, and for want of a few feet in diameter the horse mill was virtually useless. As a result the tower was turned into a store-room, a purpose which its stump still fulfils today.

The next stage of pumping, from about 1720, was in a larger and better-designed building called the Square Horse Works next to the mill tower, but pumping with horses was still very costly and the New River directors were always interested in new sources of power. By the 1760s the earliest coal-burning Newcomen type atmospheric beam engines known as 'fire engines' were in widespread use, particularly for pumping at places like coalmines where their fuel was virtually free. At this stage Smeaton's path crossed Mylne's once again. Smeaton had carried out prize-winning research into the science of working machinery by wind and water, and he had now designed a modification of Newcomen's engine that he believed would make it cost-efficient. The New River Company asked him to advise them, and he visited the site and prepared a report on the various possibilities before recommending his own new design, to which the company agreed. This was installed in a new engine house at New River Head at the time when Mylne was first working for the company, but it turned out to be very costly to run and Mylne then turned his mind to the problem.

His solution was to build an overshot waterwheel in a deep pit at New River Head, an option that had been considered but rejected by his predecessors. It was turned by water from the Round Pond, and although that water lost some altitude in the process it was still high enough to supply houses in Clerkenwell so it was a source of power that cost the company nothing to run. Meanwhile the turning wheel was geared to pumps that forced water to the Upper Pond, thus reducing the amount of work for the steam engine and saving coal, and royalties. The wheel began work in 1779 and the directors were so pleased with its performance that they awarded Mylne a bonus, 'as a present for Making Water Wheel' as he wrote in his diary. They were right to do so, and although the company in due course had a succession of steam engines from Boulton and Watt to replace Smeaton's, the waterwheel continued in use for some seventy years until about 1850. It is an example of how versatile Mylne's skills were, and is interesting because it makes a change from the usual sequence. Normally traditional power sources such as wind, water and horses are tried before newer ones like steam, oil and electricity, but at New River Head the company had used Smeaton's coal-fired engine for years before Mylne's waterwheel.[26]

In this context one action that typifies Mylne's awareness of the importance of potentially historic documents is his diary note that he 'Bought a book and Pasted the Plans of the fire Engine in it for N River Company'. Unfortunately that book does not seem to have survived, and nor has any drawing of his waterwheel.[27]

During this period Mylne took on another appointment, as Clerk of Works to the Royal Hospital for Seamen at Greenwich from November 1775. Like the surveyorship at St Paul's his acceptance can hardly be explained on the grounds that it was profitable – it brought him less than five shillings a day – nor its status, as it put him in a subordinate position to another architect, James 'Athenian' Stuart, who was already the hospital's Surveyor. The Earl of Sandwich, First Lord of the Admiralty, chaired the meeting that made the appointment and the minutes show that it was he who recommended Mylne to fill the vacancy caused by the previous incumbent's death, on the basis of his 'well-known' architectural abilities at Blackfriars and elsewhere. Sandwich is one of those politicians whose reputation now stands much higher as a result of modern research than it did in his lifetime when his integrity made enemies. A naval career with the chance of wealth from prize money was an attractive option for the portionless younger sons of good families, and accelerated promotion, which meant a larger share of any prize ship captured, was traditionally available to those with parliamentary connections or other influence. Sandwich operated a system of promotion based only on merit that made him some powerful enemies. He also introduced a number of reforms to overcome the entrenched dockyard practices and other traditions that kept the Navy constantly short of ships. Mylne seems to have been recommended to him by Joseph Banks and Sandwich probably liked what he heard of Mylne's character. Sandwich was a director of the East India Company, in which Mylne held voting stock, and this may have been a factor for Sandwich, who wielded political power from his influence in Company matters, and he certainly asked for Mylne's voting support later.[28]

For Mylne, the post had two obvious attractions. One was that Wren, whom he greatly admired, had been the hospital's first architect although he did not live to complete the work, and the appointment would give him a chance to study Wren's work while making some needed improvements. Another was that it brought an apartment for his use, on the ground floor of the east wing of Queen Mary's building. A retreat near London in a fine set of Thames-side buildings will have had its appeal to a man who was by this time the father of four young children. He paid his predecessor's heirs handsomely for the furniture in the apartment, added to it, installed a dairy cow for his children in the stable, and moved down there with his family for a few months every summer for as long as he held the post, as well as making regular working visits at other times.

As for Stuart the surveyor, twenty years his senior, he was already in failing health from his convivial lifestyle and Mylne may have foreseen early advancement as had happened at the New River Company. One disadvantage, but one that may not have seemed important at the time, was that while Stuart's post was established and gave him a seat and vote on the hospital board, the Clerkship of Works, though permanent in practice, technically ranked as a temporary post and did not carry the right to sit at board meetings. Something else he may not have appreciated was that there were already internal stresses within the institution working towards an upheaval. Unaware of the turbulence ahead, he accepted the post.

The Royal Hospital had been endowed by Queen Mary, partly as a charity for disabled seamen but also as a political statement. The French had shown, with the magnificent Hôtel des Invalides of 1671–76, how a state could express its support for its worn-out soldiers, and Greenwich provided England's answer, deliberately grander than Chelsea Hospital. At Greenwich was a superb riverside location, at the end of a long southward loop in the river. Many of London's overseas visitors still arrived by sea, and the sailing ships that carried them upstream had a lingering view of this part of the shore. It was the ideal location for a showpiece to let the world see the value Britain placed on its seamen. Queen Mary already had a house there, set well back from the river, and she insisted that it must be left with a river view. Wren's solution was to design ranges of symmetrical buildings to the left and right of an open central axis that gave a view from the Queen's House to the river and vice versa, thus benefiting both.

A French visitor gives a flavour of the place in the early 1770s:

> The hospital ... on one side the Thames, which, by its winding forms an elbow ... separated from the river only by an iron balustrade. There, vessels of all sizes, and from all parts of the globe pass and repass. As the tacking about, occasioned by the elbow of the river, retards the progress of the vessel, the sailors avail themselves of this circumstance to bring news to the invalids, or to ask them different questions, this conversation lasts as long as they are within hearing ... and is animated by that effusion of heart, so usual among old comrades, who have been a long time absent from each other, or who think this interview may be their last.

He noted that the pensioners had clean shirts twice a week, received some pocket money, and in their dining room had each a locker for his napkin and cutlery.[29] The pensioners were also entitled to a substantial diet. Each day a man received a pound of bread and half a gallon of beer. In addition there was a weekly allowance of three pounds of beef, two pounds of mutton, two and a quarter pounds of cheese and a pint of peas. Pocket money varied from a shilling to half-a-crown a week, according to seniority, and clothing was supplied.[30]

Once in this part-time post, Mylne found a range of useful tasks that needed doing. He organised some necessary building works, such as removing the inelegant pediment known as Ripley's Saddle, and cleared away a clutter of artificers' sheds that was spoiling the appearance of Wren's buildings. Learning that the convicts imprisoned on the hulks moored off Greenwich were kept busy raising ballast from the river bed while they awaited transportation, he persuaded their supervisors that the hospital should benefit from their efforts. As a result he obtained 3,000 loads of gravel for the hospital free of charge, and used it to make a network of gravel paths around the buildings for the pensioners' use.[31] In a characteristic action he gathered together what he realised were important architectural drawings by Inigo Jones and others which were lying around at risk of loss or damage, and he had these bound into three large books with some of his own drawings. Thus he spent his time usefully, and with no obvious problems at first, while the bulk of his working life continued to be filled with the New River, St Paul's, the continuing works around Blackfriars and his busy private practice.

The Governor of Greenwich Hospital being non-resident, the daily running was in the hands of the Lieutenant Governor, Captain Baillie, who had an apartment and a private garden within the grounds. An institution like the hospital was obviously open to abuse, and it seems that there was some malpractice with regard to the supply of meat and other provisions. Baillie began to concern himself with these matters and in 1778 circulated an anonymous printed attack on the hospital's management, which he soon after admitted was his work. It was aimed at all aspects of the hospital's administration and attacked officers up to and including the Earl of Sandwich – who had been tipped off about its impending appearance by another officer at the hospital.[32] Inevitably it made Baillie a popular hero who appeared to be standing up for the ordinary seamen against the privileges of rank. This had not always been his stance – the previous year he had written privately to Sandwich complaining that some proposed changes at the hospital would deprive him of part of his garden and 'expose my family upon the most necessary occasions to every creature passing and repassing'.[33]

It may be that Baillie was backed by those with a darker political agenda, for according to a contemporary historian there was 'a grand system for clogging the wheels of government' that year on the part of the opposition. Much parliamentary time that session was taken up with attacks of Sandwich's performance of his duties by Fox who proposed his permanent dismissal, by Lord Bristol who claimed the navy was shamefully mismanaged, and then when the Duke of Richmond in the Lords took up cudgels in the matter of Greenwich Hospital.[34] It was certainly a very disruptive procedure for Sandwich and many others, as it can take much time and energy to deal with even the most unsubstantiated attack. By the end of it Baillie was publishing more printed material in a strikingly costly format at a time when he was petitioning for financial help having been dismissed his post, and this too suggests he was the tool of others.

Baillie claimed that too many landsmen had offices that should have been filled by seamen. Furthermore, thousands of pounds were unnecessarily spent every year on building works even though he thought the hospital was already sufficiently grand. Meanwhile the pensioners were served bad beef and watered beer.

Annoyingly for Mylne, Baillie's technique was to attack every possible target, and this included Mylne, who he said was only there because of John Paterson's influence with Sandwich. He said Mylne was occupying space that could be used for pensioners and had turned out a butler's mate formerly lodged there to give his family more room. He had made unnecessary changes including removing some safety railings in the grounds out of mere caprice, to the great distress of blind and infirm pensioners. He kept a gate that led to the hospital wharf locked. He had made a cowhouse for his family, and the pensioners were at risk from 'droves of horned cattle'. Furthermore he was arrogant, and had 'assumed to himself the authority of Governor'. As it turned out these were all matters Mylne could easily explain or disprove, but far more irksome was an imputation that since someone of his eminence cannot have taken the job for its small salary, he must have done so in the hope of enrichment by other means. However unfounded, it was the sort of allegation which is easy to make and almost impossible to refute, and Mylne who prided himself on his integrity could not ignore it. Mylne responded as he always did to anything he saw as an attack on his honour, in great detail, and in the first instance to Sandwich:

My Lord,

As Mr Baillie, the present Lieut't Governor of Greenwich Hospital, has been pleased to set my conduct and character; as Clerk of the Works, in an odious and criminal point of view ... by means of a Book which for other base purposes, he has composed in the most virulent and false terms, ... Very soon after being entrusted, with the troublesome Dutys of that Office, I found Mr Baillie, a turbulent restless man; and by brow-beating some and teazing others, he was constantly putting the Hospital to a great expence, for Works of some kind or other, to be done to his Apartments; or by procuring allowances being made, for works he had done of himself, and getting the Board of Directors to order the Bills of such Works to be stated in the Books of the Hospital Expences. – Sometimes, when he could not procure these things in a direct Manner, he was mean enough to take others, not altogether consistent with the views of a Man of Honour.

In order to see how much Expence, he had really put the Hospital to, first and last, I caused an Extract to be made, of all the Moneys paid, on Account of the Repairs, Alterations, and Works of different kinds, which came under my department, separate from that of the Stewards or any other. I was astonished to find it little short of £1200, about this time twelvemonth. As this fact was mentioned by some of the Directors, on occasion of another of his frivolous Aplications, and as he positively denied that such a computation could be true, it was given to the Clerk of the Cheque to examine it; and hereunto annexed.

Mylne went on to explain how Baillie on another occasion obtained approval for the erection of drying posts for his washing at the hospital's expense, but then persuaded the workmen to use the timber allocated to form an 'Alley in his Garden ... to carry plants or Vines to make a shady walk.' Mylne had reported this to the Board in May 1777, several months before Baillie's book appeared. The board adjourned to inspect the posts and minuted their approval of Mylne for having reported the abuse to them. Mylne had their order copied out and prominently displayed in his office as a public warning to others that not even the Lieutenant Governor was above the rules, and this may not have endeared him to Baillie.[35]

Mylne's letter continued:

It was evidently a fraud; and had the whole Structure in Question, been pulled down, it would have been a more effectual method, of holding up this example, against such irregularities in the future ... These Transactions he has not been pleased to mention among the frauds and abuses in Greenwich Hospital, but which, I believe, are the real motives for his having paid so much attention to me, in his late malevolent publication; and will serve to shew, what credit is to be given to a man, who being guilty of such conduct himself, meanly suspects every one around him.[36]

The affair of the shady walk provides an insight into the way Mylne carried out his duties, as well as a reason for Baillie's anger towards him. As soon as Mylne discovered that Baillie had transmuted clothes poles into arbours by telling the carpenters to ignore Mylne's orders and follow his, he wrote to the board setting out what had happened and explaining his frustration with the problems such behaviour caused:

> Workmen in general wish for no other, than variations from the real Orders, for the purpose of eluding the proper investigation of what they execute, and perplexing the charge thereof to be made. Very early after entering on my Duty I wrote circular Letters to all the workmen, enjoining them not to listen to the directions of any Officer, in performing your orders, which I had to retail to each – yet for all that precaution, and my express instructions at all times, I cannot get any of them to perform a piece of work exactly as it is ordered; and sometimes it is stopt, delayed or altered … Estimates and cheques are of no use and avail nothing – This particular business in the conduct of a Superior Officer, is of bad example to others.

Mylne was not being merely officious in reporting Baillie to the Board. Not only did he have a plain duty to report the matter, some new hospital regulations made the previous year defined the duties and responsibilities of all the officers, in an effort to prevent precisely the kind of abuse Baillie would later complain about. Those concerning the Clerk of Works spelt out that if he permitted any works or repairs that had not been authorised by the Board, he would be personally liable and the cost could be deducted from his salary.[37]

Inevitably there was an internal enquiry into Baillie's allegations. One of the Adam brothers came down at Mylne's request and gave evidence of his warm approval of the work Mylne had done there. None of the charges was proved, and the board reported that so far as Mylne was concerned, he appeared to have been in every respect a good servant of the hospital. Overall they found that Baillie's allegations were malicious and void of foundation. How he should be dealt with was postponed until pending legal proceedings were finished, but in the meantime they suspended him from office.

The first response of some of those attacked in Baillie's pamphlet, including Mylne, had been to seek redress through the courts alleging libel, but in November 1778 the court of King's Bench decided that it could not hear the matter because it was really a dispute between the hospital and Baillie. He was one of the directors, and the board had already dealt with him internally by holding an enquiry. A further legal difficulty was whether Baillie had himself 'published' his pamphlet in the sense required for libel if he had merely supplied it to fellow directors and others had disseminated it more widely. Baillie's affidavit in the case had backed down from his earlier attack on Mylne, and said that he had never meant to accuse him personally of any wrongdoing. However he still claimed that Mylne was seeking to add unnecessary magnificence to a building which he claimed was already finished, and said he had shown '... an overbearing and factious disposition by taking a warm and active part in all disputes in so short a time after his being appointed'. It also complained that Mylne had described Baillie as a blackguard to the committee of enquiry and suggested that he was trying to turn an occasional job into a permanent one for the advancement of his own career. Lord Mansfield, who gave the judgment, said that clearly amounted to an insinuation against Mylne. The court made no finding that any of Baillie's allegations were well founded, and conceded they were potentially libellous, but merely decided it had no jurisdiction in the matter. Despite this, Baillie's supporters trumpeted the discharge as being a vindication of all his allegations. Those attacked countered with their own pamphlet, setting out the official findings of the enquiry in detail, and pointing out that Baillie's claimed success in the libel suit was nothing of the sort.[38]

All of this left Mylne among others feeling very dissatisfied with the outcome, and the commissioners and governors, including Stuart, petitioned the Admiralty that Baillie must be dismissed, with a supporting petition signed by others affected including Mylne, who also wrote directly to Sandwich on 5 December. For twenty years past, he said, he had been engaged in conspicuous public works, without the slightest suggestion of any impropriety. Now, within three years of going to Greenwich, where he had been far from idle, Baillie's book made him out to be a 'thorough-paced scoundrel'. Now he was being daily made aware of the effects of what he called the malicious madman's writings. Like many other of the officers, his future fortune and bread depended upon having an unsullied character. Every day more of the public heard and believed the allegations, to his detriment. The Hospital must deny it and the Admiralty Board should assist. He expressed gratitude for the committee of enquiry that was to follow, trusted it would wipe away every stain thrown at him, and said he would cooperate in any way that would assist them. The board should know that he had already taken legal advice and had applied to the Court of King's Bench to obtain redress, but they took the view that the Admiralty had jurisdiction over Baillie, even though the Chief Justice expressed the view that the book contained libellous calumnies. Baillie, said Mylne, should be dismissed.[39]

All this must have had its effect, and the hospital commissioners wrote to Baillie on 11 December to say he was dismissed with effect from the next quarter day, Christmas Day 1778.

The loss of their champion was certainly a disappointment to Baillie's supporters, and it may not be a coincidence that a few days later, on the night of New Year's Day 1779, a mysterious fire caused extensive damage to the chapel and adjacent parts of the buildings. Mylne, summoned from London to help, rushed down and directed some expedient demolition to halt the spread of the fire. Sir

John Fielding the almost blind magistrate who had succeeded his half-brother Henry, the author of *Tom Jones*, at Bow Street was asked to conduct what he called:

> … as elaborate an enquiry as ever I made in my life… for it was represented to me … it might have been done maliciously … we went further and pursued every little hearsay; of any old nurse, or any drunken pensioner …

Suspicion fell on a tailor's room within the premises where uniform suits were made, and beneath which the fire had started. It is clear from the evidence that it was not the workplace of craftsmen but a sweatshop where semi-skilled men toiled with needle and thread. According to one ex-employee, Thomas Bird, who ran it kept a cask of gin in the room and dispensed it on request, as well as beer, stopping the cost from the men's wages each Saturday – so that in some cases there were no wages left. Three weeks earlier forty men had drunk five gallons in a day. One Wilkinson had drunk thirty-three glasses of gin in a day. People smoked as they worked, and there were lit candles. Against this picture of tipsy disorder, Bird in his turn claimed nobody smoked except one man he had discharged a month earlier. True, there was a fire in the grate, but he had been most careful to have it extinguished with three pitchers of water by one of his workers before locking up and going home on the evening before the fire. The man he named said it had been two pitchers, but otherwise confirmed this.

Fielding concluded the fire had not started in the tailor's shop as alleged but in the structure underneath it. 'In order that we might investigate this matter with exactness', he wrote 'Mr Mylne, the Surveyor there, a very ingenious man, as I could not see objects so well, he made a model of the under part of the floor, covering over the altar piece, where the fire evidently was seen.'

As a result the cause of the fire was never established, while Baillie refused to accept his dismissal, and remained for a time in his apartment.

By early 1779 Sandwich's supporters had produced their own pamphlet. The authors said they were concerned to put the record straight, given that 'the Public in general … are so easily induced to believe that all public Business is improperly conducted' and might think the greater part of Baillie's allegations must be true. Before setting out a summary of the internal committee's findings, followed by a 'few genuine anecdotes' about the ways in which Baillie sought to profit from his office, the pamphlet castigated Baillie and his presumed backers. The author, it suggested, was clearly a man of genius and therefore plainly not Baillie – he was just the labourer who had:

> … furnished the Materials, raked every Kennel (i.e. gutter), and ransacked every Hole and Corner for them … so far indeed he may take to himself the Credit of the Performance; but he cannot be surprized that a Work composed of such

> materials should, the Moment it became exposed to the Sun-shine of Truth, crumble into Atoms, and bury its projector in the Ruins.
>
> How can he, without a Blush, pretend to be a *Friend* to the *poor Pensioners*, when it is notorious that he has treated them with the most rigorous Severity? Does he forget the Time when he was the only Officer ... who, for Winters together, kept those infirm debilitated old Men standing Centinels at his Door the whole live-long Day, among the Severities of Frost and Snow, in a Passage frequently streaming with Water, and exposed to the Torrents of a North-east wind? Would a Man of common Humanity have called forth the poor worn-out Saviour of his Country, from the Fire-side which the Bounty of that Country had provided for him, to shiver for Hours together at the Threshold of his Door, where the Eye of Compassion must have melted at seeing even the Stranger's Dog waiting for Admission in vain?[40]

The House of Lords then held a committee of enquiry into the management of Greenwich Hospital. It began on 11 March and ended on 8 June 1779, and this time-consuming procedure may have been the whole purpose of those behind Baillie's allegations. During this time, an unconnected personal tragedy befell Sandwich with the death of Martha Ray, his mistress of eighteen years who had borne him nine children. Sandwich was a leading figure in the revival of interest in 'ancient music', while she was said to have perhaps the best singing voice in England. They lived together openly and she acted as his hostess in London and at his country home at Huntingdon. She had had a brief affair with James Hackman, a young army officer Sandwich had invited to Huntingdon. The Tahitian Prince Omai, brought to England by Captain Cook, saw them together and though unable to speak English, managed to convey his suspicions to Sandwich, who asked Ray about it. She admitted the affair and promised to end it, while Hackman left the army and took holy orders but continued to be obsessed with her. On 7 April 1779, wild with jealousy, he followed her when she left the Admiralty alone in Sandwich's carriage, and later shot her through the head as she left Covent Garden Theatre after attending a play, then tried to kill himself with a second pistol. Sandwich who had been waiting at home for her to return for a late supper rushed upstairs and threw himself on a bed when he heard the news, saying he could have borne anything but that. Hackman was convicted of the murder the following week and hanged three days after that. Before his execution he wrote to beg Sandwich's forgiveness, who replied that he forgave him, but said the death had robbed him of all his comfort in this world.[41]

Less than a month later at the House of Lords enquiry Sandwich defended himself robustly, in a speech that treated Baillie's complaint as the disguised attack it was on his own position. Quoting from the pamphlet he said he stood accused of turning the hospital into an engine of political corruption, of perverting the Government and design of the hospital and treating it as an appendage to his private fortune, yet none of that could be shown.

As to the complaint that landmen were appointed to office, he ridiculed the idea that all the offices could be filled by disabled seamen. The surveyor and clerk of works must be eminent architects, while the organist had to be a musician, not a one-armed seaman. To argue otherwise was like saying that because Bedlam was a hospital for lunatics, its managers must be madmen. If some of the office-holders had lodgings in the hospital, why should they not? It meant they were close at hand to perform their duties and saved the expense of house rent that would have to be paid for them if they lodged outside. It was not as if they were taking up space that would otherwise be used for the pensioners, for the house had heard evidence that the number of pensioners was limited not by any lack of space but only by a lack of greater income for their support.[42]

Sandwich was able to show that during his time the hospital had undergone a major expansion and now held far more pensioners than ever before – an increase of almost two thirds. Before the House of Lords the original allegations were repeated, added to, and refuted by witnesses on both sides, and none who was mentioned can have enjoyed the process. By now no detail of the hospital had escaped scrutiny, either in the various pamphlets, in affidavits, or at the internal enquiry. Pensioners had supposedly been moved to make way for the secretary's apartment. The chaplains had been drinking communion wine in their apartments. The laundry was filthy, unwholesome and came back smelling of hog dung. Pins that a pensioner hid inside his dirty shirts were still there when the laundry came back, showing that they had not been washed. The shoes and socks supplied were of poor quality and the bed linen undersized, particularly that supplied by the present steward. The drains were bad. The secretary was a great nuisance. The beer was watery and dreadful. The butcher supplied pork from old sows – a pensioner had seen the long dugs on a carcase. Other meat had been stolen, or was hard and of a disagreeable character. A lieutenant said the chaplain had warned him not to support Baillie, at the risk of it stopping his preferment, while a pensioner claimed the chaplain had threatened to disinter his wife's body because he disapproved of the burial place. Another complained of landmen with sumptuous and elegant apartments, better than their naval counterparts.

Some of these charges were soon dealt with. The reason for the watery beer was that one of the pensioners who distributed it watered it to conceal the large amount he had embezzled and supplied to those who were prepared to reward him with money and drams, for which he had been expelled when detected. What was more, it seemed he was now among Baillie's supporters and it was Mylne who had detected the fraud.

There were lighter moments. One day in March there was a demonstration of Baillie's supporters, including male pensioners and their ladies who lived outside the hospital, cheering and crowding their lordships as they went in. The Lord Chancellor thought it very improper and unseemly. Lord Fortescue told the house that he had never been so kissed in his life and that it was unfair of the women to attack him in that manner when he was far too old for a suitable response, while other peers thought that every British subject had the right to express his views at the doors of Parliament.

Mylne's supposed crimes did not occupy much of their Lordships' time, but were referred to. Captain Allwright complained that the chimney pots on some apartments near his had been changed, forcing smoke down his chimney, and Mylne had not done much about it. Mylne cluttered up one shed with his coach and horses, and kept two cows. He had refused to open a gate for a timber-cart. His labourer had been stealing hospital bricks.

Mylne's supposed high-handedness was mentioned again. He had come to blows with one of the naval officers; though it then emerged the other man had followed Mylne back to his apartment after an argument so that Mylne was acting in self-defence. Captain Chadds gave evidence that he had feared the injured man might issue a challenge, and steps were taken to see that Mylne and the lieutenant shook hands, and they had remained friends ever since. A pensioner who had been dismissed from the hospital at first claimed it might have been because he kicked Mylne's dog out of the painted hall, whereupon Mylne had shaken his fist at him, called him a son of a bitch and demanded to know his name, but then admitted he was dismissed after refusing to accept a punishment the court of governors had imposed on him for a completely different matter.[43]

In all this it is difficult to find any allegation of substance against Mylne. He certainly kept a cow, in a small stable used by some predecessor for a saddle-horse, but he pointed out that he had four children under the age of six and 'cannot do with the adulterated milk of Greenwich, no more than that of London.' Yes he had moved a number of workmen with objectionable trades away from the immediate vicinity of the hospital – a smelly oil-man, a noisy stonecutter, an armourer whose smoky forge had been within 15ft of the hospital's white stonework, and a plasterer and bricklayer who poisoned people with their new-slaked lime. All this nuisance was removed at a cost of less than £2 a year in lost rent. As to his extending the gravel walks, this made use of ground where previously 'nothing but publick nuisances of shitting, pissing, Mending Boats, Boiling Pitch Kettles ... served to entertain the Eye and regale the nose of all Company on the Terrace Walk.'

Eventually the House of Lords, by a vote of 67 to 25, decided that Baillie's allegations were groundless and malicious as regards Sandwich himself and also as regards the commissioners, directors and officers of Greenwich Hospital. As in all such cases there were still doubtless many who believed the allegations. Baillie remained dismissed, but less than three years later the Duke of Richmond, who had fostered his complaint, was in the cabinet after a change of Government and appointed Baillie to a lucrative post for life under the master general of ordnance.

By this time major rebuilding works were necessary because of the fire, and could only proceed if Stuart and Mylne could work harmoniously together, as each had a part to play. For a time they managed, but eventually the strain began to show. Until now Mylne had been able to find ways of doing what was necessary, making his own drawings where Stuart failed to do so, but this became impossible where a major rebuilding operation was needed. The likelihood of a collision between Stuart's fuddled indolence and Mylne's incisive energy was now very real. The distinction between their roles meant that it was Stuart's job to provide drawings for the restoration and Mylne's to oversee the way it was done, but while Mylne was anxious to restore the building 'to its former splendour,' Stuart seemed incapable of any urgency. Mylne wrote repeatedly to Sandwich asking him to visit Greenwich to see what was happening, and complaining of parsimony 'in the absence of the Surveyor (who is at Bath for his health)'.[44]

A surviving note in Mylne's hand shows what he needed. It is headed 'List of drawings necessary to be made for the reconstruction of the chapple & cupola.' Twelve plans and sections were listed, which were to be at a scale not less than 3in to 10ft, and all to the same scale. Dimensions must be shown in ink, and there must be letters, keyed in the margins, to show which parts were to be of stone, marble, oak, fir, mahogany, stucco, wainscoting, painted or gilt. Put as clearly as that, one can well imagine the fertile room for confusion a mischievous contractor might cause if asked to price the work on the basis of anything less exact.

Mylne also continued to refuse to approve any work that he considered an unnecessary diversion of the limited hospital funds that were needed for rebuilding the chapel and linked works, and this built up a fund of resentment in the directors and officers affected. Thus the Taylor family could not have some unnecessary wallpapering in their apartment, nor could Mr Maule have some work he wanted. In December 1779 Mylne wrote to the board explaining the constant struggle to avoid useless expense, so that the available money could be laid out on the 'great and good Works' of benefit to the hospital. At the same time he carried out various improvements to the wards used by the pensioners so that more could be accommodated, and made better storage arrangements for the coal supply so that it could not be pilfered so easily. As well as building works Mylne addressed the hospital's water supply, asserting forgotten rights to certain springs and thereby reducing the amount that had to be purchased from the Deptford waterworks. These were all time-consuming exercises and it is plain that he took his responsibilities seriously.

That was the position as the 1770s came to an end, but the malevolence at Greenwich was not finished with Mylne yet.

9

'... a man of pure honour...'

In 1780 Mylne was forty-seven and his family was still growing. He already had four daughters and a son, Robert. The following year William was born, and in 1782 Thomas, who died before he was six months old, possibly of smallpox. Vaccination was becoming popular, and in November 1782 Mylne mentioned in a letter to a friend in Paris that all seven of his children had been inoculated but that he had lost one.

The decade began with civil unrest in London, with the Gordon riots in July 1780. Ostensibly an anti-catholic protest, all the disaffected of the city eagerly seized an opportunity to break the usual rules. Newgate and other prisons were burnt and their inmates freed, and many houses burnt to the ground. Mylne found himself affected on two fronts, the New River being the first. It was believed that the rioters planned to set the Bank of England on fire and cut the New River to hinder fire fighting. As a result troops were stationed at New River Head and other vulnerable points up its course such as the wooden aqueducts at Highbury and Bush Hill. The attack never materialised but the soldiers proved reluctant to leave. The company had ordered that the officers should be given a two-course meal with wine each day, and though the rioting was over in June and the food and wine discontinued in October after earlier hints were ignored, some of the military were still there at Christmas.[1]

At Blackfriars Bridge there was serious rioting. The toll had never been popular – despite the fact that the bridge replaced less reliable and less convenient ferries that made a similar charge. When the mob crossed the bridge they took the opportunity not only to loot the tollhouses, escaping with 'bushels of halfpence' according to Walpole, but also set them on fire. Later, soldiers dispersed the rioters at the bridge and several were pushed over the balustrades into the Thames.[2]

It must have been galling for Mylne to see his attractive little classical tollhouses destroyed, but he set to work and soon had them rebuilt. They were of wood rather than stone, because the toll was always intended to be temporary until the bridge was paid for, and as a result they proved an easy target.

After this episode Mylne's life soon resumed its normal pattern. He designed and built a new bridge at Romsey, designed another for Londonderry and advised on one over the Menai straits, though none was built there until years later. He gave evidence on the reasons for the collapse of Smeaton's bridge at Hexham. He advised on copper mines in Anglesey, and gave evidence to the House of Commons about the working methods and the wind- and horse-driven machinery used there for raising water. He fitted lightning conductors to St Paul's, with advice from Benjamin Franklin, and reinforced the structure with concealed ironwork to prevent further settlement of the dome, work so extensive that the cathedral was closed for two years. In 1789 he arranged the cathedral for a major public thanksgiving service on St George's Day to celebrate George III's recovery from insanity, the first royal visit since 1715 and a national occasion. He also surveyed Rochester Cathedral and had work done there.[3]

He designed the elaborate plaster ceilings and wall decorations for the best rooms at Inveraray Castle and major works at Rosneath Castle. In Edinburgh he advised the corporation on its water supply, partly built by an earlier Mylne. He was asked to provide plans for a new market and slaughterhouse there, and showed his regard for his birthplace by charging a nominal half-guinea for a large plan, two elevations and four sections all neatly washed, drawn and figured, and sent up by the post coach in their own tin case.[4] He planned to transfer the estate at Powderhall to young Robert in due course, and continued a programme of improvements there, planting an avenue of elm and holly and building a lodge. He also designed a new house to replace the old one, though it was never built, and made notes of the good views the upper rooms would have over the Firth of Forth to Burntisland. He made notes for converting an old coach house there into a gardener's cottage, and for the entry to a cellar to become a grotto embellished with 'Gothick ornaments from Mrs Coade'.[5] When the tolls at Blackfriars ended in 1785 he bought one of the old tollgates for ten guineas and paid to have it sent four hundred miles to be hung at Powderhall, presumably for sentimental rather than practical reasons.[6]

On the New River he arranged for one of Boulton and Watt's steam engines to follow Smeaton's, and replaced a leaky wooden aqueduct near Enfield dating back to 1612 with the long clay embankment still in service today.

There was a brief addition to the Mylne household in 1783, in the form of Robert's nephew Willy, the illegitimate son William had left in his sister's care when he fled from Edinburgh. Anne had seen to his schooling, paid for by William when he could raise the funds, so Willy had studied Latin at the High School and then went to Sedbergh. His father once hoped to set him up in the American colonies, writing to Anne 'this is the country for such as him that labours under the stigma of bastard', but his return to Britain had ended that idea. Now Willy came to London needing a career, and the decision was made to send him to India, in the East India Company's artillery. This may have been helped by Robert's holding of East India stock – there was keen competition for all such vacancies.[7]

As William was still in Dublin Robert kitted him out, and noted the details in his diary. Apart from his uniform and spare clothing Willy needed a short sword, a brace of pistols, a fusee, military books, maps of Bengal, a case of instruments, a hammock and camp stool, a watch and chain, a trunk and liquor case, £40 for his messing on shipboard to India, with a further £20 sterling, £12 in Portuguese dollars, and ten pounds of tobacco. Once he was out there a further box of books and instruments followed, and it is clear that Robert did his best to make sure that Willy went out as a gentleman and had a good start. His father was to pay the bill when he could manage it, but he never seemed to have much to spare and paid only half that year. Willy was duly given a cadetship and arrived in India in September, reaching the rank of Lieutenant Fireworker in the Company's Artillery at Prince of Wales Island by 1787.[8]

The older William still ran Dublin's water, but he visited Robert in September 1787, and the brothers went off for a month on their first major trip for thirty years. Robert had been asked to see some of the naval installations at Portsmouth and they clambered over warships including the 74-gun *Edgar*, of which their niece's husband Charles Thompson was Captain, and inspected the dockyard and fortifications. They then spent some time in Bath, and went on to Ilchester, Wells, Bristol and Shropshire, stopping along the way to visit some of Robert's clients, before returning for another few days in Bath – Robert suffered from rheumatism – before going back to London, where Mylne soon wrote to Matthew Boulton, by now a friend, asking a favour for Captain Thompson. For some years past the great guns of British ships had been increasingly fitted with flintlocks to provide a spark to fire the gunpowder, instead of the traditional smouldering fuse. Using a flintlock operated by a trigger allowed each gun-captain to stand behind his gun instead of beside it and to take careful aim along it before pulling a cord to fire it at precisely the right point of the ship's roll. Thompson had this mechanism on his guns but he wanted to add a second flintlock to the left-hand side of each to reduce the number of occasions when a shot was missed because the flint failed to spark correctly. The dockyard could supply right-hand ones, used for pistols and muskets, but not the eighty left-handers he wanted, and Mylne asked Boulton to seek them from a Birmingham gunsmith.[9]

Robert was still Clerk of Works at Greenwich Hospital, and probably thought his troubles at an end after the House of Lords enquiry. After the fire he had given up his apartment to make space for displaced pensioners and he rented a nearby house instead. His 1780 diary shows him paying bills for plants and seeds for the garden there and he moved his family down for July and August, so it seems that he expected to stay.

Within a year came the next problems at Greenwich, when a series of anonymous letters signed 'Zero' was sent to the board.[10] They made new allegations against Mylne, that he had favoured friends from the Blackfriars Bridge committee when ordering materials for Greenwich, had plastered 'Scotch mud' on the colonnade and had wasted hospital money on repairs. The directors ordered that the letter should be sent to Stuart for him to consider and report, and the matter was then repeatedly postponed until he reported that the allegations were groundless and in some cases quite false.[11] Nobody from the bridge committee was supplying materials. The 'Scotch mud' was the stucco recently invented by the Adam brothers, and had been properly used by Mylne at Greenwich to make a very economical repair to some fire-damaged stonework. It seems clear that someone wanted Mylne to be dismissed, and the likeliest candidates are among the contractors who had been accustomed to a more easy-going and lucrative regime before his appointment.

Around this time, in May 1781, Stuart and Mylne had jointly proposed to the board that the craftsmen from Somerset House 'who are the first of their respective professions' should be employed reconstructing the chapel & cupola, and paid on the same measured basis as at Somerset House. This sounds like Mylne's suggestion, and was not popular with those accustomed to work at Greenwich, so the decision was deferred. Perhaps earlier gossip about that proposal provoked the anonymous letters. In June Mylne wrote, and Stuart countersigned, a letter reporting on the progress of work. On 4 July 1781 a general court of the hospital governors agreed that, because of the greatly increased workload resulting from the fire, Mylne should receive a suitable extra payment when the work was finished.[12] In August he proposed that a great improvement could be made to the 'light, air and chearfulness' of the upper wards in Queen Anne's Building if the window breasts were cut down to floor level, and this was approved.

By December Zero was writing again, apparently claiming that Mylne had signed agreements with the new workmen for a share of their bills, and on 8 December the directors resolved to advertise:

> ... three times in each of the morning papers of the greatest note, that if the Author of that letter would attend the Board [or any of them] and make good what is therein set forth he shall have all fitting encouragement and his name kept secret if desired.

The same day Mylne received something of a rebuke for having failed to be present when some measuring was to be done – 'in consequence of anonymous information which had come before the board' – he accepted he had not been present but said he had made the necessary checks afterwards.

A letter of 15 December 1781 from Mylne deals with this allegation. He had followed hospital rules as much as anyone could. His workload had greatly increased since the fire, and he had done much work that was properly Stuart's duty. 'It will not be improper to add, that in the half-dozen of years I have served Greenwich Hospital, I may almost literally say I have not got a shilling from it, but what I have spent and paid away, not for myself but for the good of it, and the due performance of its business.' Then on 16 December Zero wrote again, now backing down from his assertion that Mylne had signed agreements with the new tradesmen but claiming they had agreed to pay two and a half per cent for 'contingent expences'. Zero claimed that his rank in life placed him above the rank of workmen, and his information came from a very capable and judicious surveyor whom he claimed he had paid to mix with the new tradesmen and discover what they said. Why he had chosen to put himself to such trouble and expense he did not explain. The board ordered Stuart to enquire among the tradesmen whether Mylne or anyone else from the hospital had ever solicited them for payments. Predictably no such evidence emerged, but even groundless allegations are damaging. The idea of taking bribes from workmen seems out of character, and this is confirmed by a minute made by the New River directors four years earlier. After Mylne had reduced a contractor's claim for some brickwork done the contractor sent him a letter enclosing five guineas as an inducement to approve the full amount. Mylne told the board, who thanked him and directed the man should have no further work from the company. In all Mylne's decades of service at the New River the only surviving complaint seems to be an anonymous note signed 'a well-wisher to all Honest Englishmen' complaining 'Before the scotchman came you were so good as to allow the carpenter and bricklayer 2 pints of beer a day but now they have Lost it – who has found it I cannot tell'.[13] If Mylne had been corrupt there would have been stronger allegations than that.

During 1781 Sandwich had two anonymous letters about Mylne. One was signed 'Timothy Stoutsoul' and the other merely 'your lordship's true friend', but though they are in different handwriting both used a similar turn of phrase that hints at a common origin. Taken together, these foresaw that Stuart would soon have to be replaced and claimed that Mylne, who would doubtless presume to solicit for his job, was incompetent, had put the pensioners' lives at risk by loading the building with brickwork it was never meant to support, had a 'consummate and ungovernable obstinacy', had insulted every officer in the hospital and was insolent and a discredit to his supporters – clearly a reference to Sandwich – and that his promotion would distress Sandwich's friends at the hospital and disgust the public at large. Was he to rise over the heads of others just because he was a proprietor of India stock? Interestingly, neither of these letters accused Mylne of any lack of probity, merely of incompetence as an architect and an abrasive manner.[14]

Then in 1782 the relationship between Mylne and Stuart reached a crisis point. On 8 January Mylne wrote to remind him that a month had passed since he had told him the workmen needed designs for the chapel and cupola. He proposed how some of the work might be done, and said that some of the necessary designs had been ascertained and some specimens done. But he reminded Stuart that since he had 'neither Orders, Directions nor Drawings, I must beg leave to refer you to the Works themselves for the Detail ... I entreat your attendance to see that things are to your mind and as they should be...'

On 4 May, Stuart wrote to the board, saying that on 12 January he had visited the chapel in Mylne's absence, to find the masons carving mouldings to a design he had never approved, and which he thought repugnant. Then Mylne had written seeking clearer instructions, but had declined to follow them when given. An examination of Mylne's letter shows that he reminded Stuart of his frequent requests for the drawings, and said he had now learnt that Stuart had countermanded work he had given the masons to prevent their being idle. He wanted to have some positive instructions before he could proceed any further.

On 10 May, it was Mylne's turn to put his complaint before the board in writing. Work on the chapel was being held up because Stuart had failed to produce drawings though he had known they were needed for more than a year, and even though Mylne had provided him with written details of all the dimensions nine months earlier. Despite repeated letters the drawings never came, so Mylne had made some himself so that the workmen would not be idle, and Stuart was told exactly what was being done. Then in January Stuart had gone directly to the workmen and told them to stop work, without telling Mylne, and when Mylne wrote asking for instructions, none came. Stuart had still not produced a single drawing of the chapel interior, the roof or the cupola. Producing the drawings was Stuart's duty: Mylne's was to execute them according to those designs, and he could do no more until the drawings appeared. Some scraps of drawings that Stuart had given direct to the masons were no substitute as they were neither clear nor intelligible. Summing up, Mylne wrote, the stone was cut, the timber was felled in the wood, every preparation was done yet the workmen were waiting. Everything was at a standstill because of Stuart's failure to do his job.

On 11 May the board found there was substance in Mylne's complaint, and ordered Stuart to provide the necessary drawings including the roof and cupola 'with all possible dispatch', to be laid before them for approval, and then to be made available so that the reconstruction could proceed. Two weeks passed, and then Mylne wrote again. Contrary to all the rules of the hospital Stuart was authorising works direct to the workmen, so that Mylne as Clerk of the Works did not know what had been authorised. It suited the workmen very well, but how could he carry out his duty to the hospital in such circumstances? His letter was read at a directors' meeting the next day, at which Stuart was present. His comments are not minuted but the board asked him and two other members to consider the matter and report back.[15]

On 6 July Mylne wrote again to Stuart, renewing his request for explanations of some drawings Stuart had by then provided. He pointed out that certain measurements differed between drawings, and what he said were problems with pedestals, column bases, and how exactly the folding entrance doors were supposed to open. Although showing some impatience the letter is civil enough, and ends 'your speedy answer will forward the works and oblige, your humble servant.' Mylne had other things on his mind just then, for that month his third son was born, named Thomas after Robert's own father. Sadly the pleasure was short-lived and though he survived long enough to be christened, he died just three months later.

On 25 July Stuart wanted his drawings back, and the board directed that he should have them, but must return them to Mylne in two weeks' time, one of the few details of his time at Greenwich that Mylne noted in his diary, presumably so that he knew when the fortnight was up. Predictably enough, on 7 August he had to write to Stuart again:

> I am under necessity of putting you in mind that the fortnight is elapsed within which the Board of Directors ordered the two drawings, which, with their permission, you had from me, should be restored. The works of the chapel stand still in several particulars for want of this. I am also to remind you that the works at both ends of the chapel, particularly at the entrance, have stood still for some time and can't be proceeded on without special & particular drawings for those purposes.

This polite letter produced an irritated answer from Stuart the next day, who clearly had no intention that Mylne should see those drawings again:

> Sir, I have repeatedly told you that you are not to work from the general drawings and I do again strictly ... forbid ... every attempt to deduce particulars, whether mouldings or dimentions from the general drawings, which therefore cannot be of any use to you ...

On 14 August a despairing Mylne wrote again to the board, sending copies of the letters that had passed between him and Stuart. He said it was impossible to complete works of such magnitude where the necessary drawings were not provided, and where scanty, self-contradictory and unsigned drawings on brown paper were given by the Surveyor directly to workmen in London and unseen by Mylne. Apart from anything else costs would rise materially. The lack of drawings for the cupola meant it was in the same unfinished state as four months ago. He had repeatedly written to the surveyor without success and repeatedly explained to the board what the difficulties were. He continued:

> I entreat in the most serious manner to know, what is to be done, what standard is to be adhered to – The Repairs of the Chapel cannot proceed under such untoward circumstances – The Orders of the Board and those of the Surveyor cannot both be followed ... If the Board does not, or cannot, procure the proper and necessary Drawings, I cannot see how these works can be proceeded upon. The Workmen must be drawn off by degrees, and the Chapel become as desolate a Scene as that of the Cupola.

A week later Mylne wrote to the hospital secretary enclosing one of Stuart's drawings as requested, and pointing out such oddities as folding doors that opened outwards over the staircase, sweeping over the top steps where the person who opened them would be standing. He also wrote that Stuart was suggesting a manner of work not authorised and likely to be a fire hazard, for example by putting in a wooden staircase. Mylne suggested it was this mixture of wood with stone that had caused the fire in the first place. Again he urged that such important work could not proceed without proper drawings, saying they could easily be done in four to six weeks, and that:

> ... nothing is now so cheap as the labours of draughtsmen... With these drawings, every business would proceed with ease and satisfaction ... without them, all will be confusion, contradiction and misunderstanding; by which much unnecessary expense will be incurred.

The same day Stuart in turn penned a long letter of complaint, from his home at Leicester Fields. In essence he said it was now impossible for him to work with Mylne any longer. The Directors had been good enough to give him copies of Mylne's complaints about him, and he gave his answers. Mylne, he claimed, had falsified the rough drawings Stuart had prepared for the masons – quite how was not explained. A sudden and violent fit of colic had prevented him from returning his general drawings to Mylne as promised. Mylne had pestered him for the drawings, saying that the work was at a standstill, he had written back warning Mylne not to make any assumptions from the general drawings. He claimed that Mylne must have altered his drawings, and accused him of a vindictive wish to spoil the chapel for the sake of ruining Stuart's reputation. Mylne had subjected him to personal insults, had devised a scheme to increase the number of beds in the hospital without consulting Stuart, and had unjustly accused him of wasting the hospital's funds. His behaviour to some of the other Officers was notorious. He was arrogant and malignant and had a fixed and rooted hatred of Stuart, who he wanted only to supplant. Stuart also complained that Mylne had not been residing at Greenwich, but usually only came down on Wednesdays and Saturdays for a few hours.

On the face of it Stuart's allegations seem very serious. It would be difficult to dismiss them now simply on the basis that they do not seem consistent with what is known of Mylne's behaviour elsewhere. Although Stuart had a reputation as a drunkard, is it conceivable that he would have been guilty of falsehood about Mylne?

The surprising answer comes from Elizabeth Montagu, a leading society hostess and a client of Stuart's, for whom he had designed a magnificent mansion in Portman Square. Just two years earlier, she had written two letters that happen to include her assessment of Stuart's character. To Matthew Boulton she wrote that she had very little assistance from Stuart, who was idle and inattentive. To her own agent Leonard Smelt she wrote that what contractors could achieve from other architects only by bribery, they got from Stuart by feeding him tobacco and pots of beer in taverns and night cellars. Sometimes he was drunk for a fortnight at a time, once so badly that she doubted he would live. More to the point, you could rely on nothing he said, as drunkenness led him into falsehood. Again and again, she wrote, she had proved this by confronting him with workmen he blamed for neglecting their orders, when he had to admit that the orders had never been given. From this letter it seems clear that Mylne's complaints were only too true. Interestingly an affidavit from a plasterer among the papers of Greenwich Hospital accuses Stuart of having taken cash bribes, rather than drink and tobacco, from him and his partner during the building of Mrs Montagu's house.[16]

Mylne, in his turn, made a long report to the Governors on 31 August. Neatly written, it spreads over seven double-sized sheets. The first part of the work needed after the fire, restoring Queen Mary's building had all been done to the Board's satisfaction. When the next stage was ready to begin, the restoration of the chapel and cupola, Stuart had failed to do his part and had merely interfered with such work as Mylne began. Some drawings finally appeared towards the end of May – far fewer than would be needed – but they did not conform to the estimates previously agreed. Not only was the design different, it was far more costly. Doing his best to cope Mylne had written to Stuart with specific queries on 28 May but received no answer for six weeks, and no proper answer even then, so that the work could not proceed. Two drawings he had returned to Stuart on the Board's instructions for a fortnight were still not returned a month later, although badly needed. Things could not proceed on such floating uncertainty. The Board in its wisdom had provided a set of rules, a constitution to guard against wasted expense and irregular acts, but Stuart would not follow it. Mylne had been ready to complete the rebuilding of the cupola since the spring, but still had no information as to what design was proposed for it. For too long Mylne had been forced to do Stuart's job as well as his own. Now it was time for decisive orders to be given, so that the ordinary rules would be followed and the work could proceed to completion.

By now the incompatibility of the two men was clear. Any protection Mylne might have hoped for from Sandwich was gone, as Sandwich had been out of office since the fall of Lord North's ministry in March 1782 following the British defeat by rebellious American colonists and French troops at Yorktown. The matter was considered at two general court meetings, on 3 and 10 September, Stuart being present at the first. It was clear from the directors' report that the two men could no longer be expected to work together. Some of the hospital's officers, as well as most of the artificers complained of Mylne's unfriendly behaviour: though it is telling that there was no hint of any corruption, and Zero's allegations had obviously proved as groundless as Baillie's. Nevertheless, Stuart was an established officer and a convivial fellow-director, Mylne an unestablished subordinate who had refused the personal requests of some of the board members and whose efforts to protect the hospital revenues had made him few friends. His fate was soon decided, and the secretary wrote to tell him that because the disagreement was stopping any progress, and having heard other complaints of his behaviour, the board had unanimously decided he should be dismissed. Stuart's assistant Newton was eagerly waiting in the wings and was appointed as Mylne's successor.

To Mylne the decision was an outrage. Was he to be punished for Stuart's failings? He was not to be cast aside so easily. When Stuart took Newton down to install him in Mylne's office, Mylne refused to leave saying that he needed time to clear his effects. On a later visit he said he was still not ready to leave, but instead set off for a planned two-day survey of Rochester Cathedral. Two days later his clerk Donaldson was in possession with orders not to leave unless forced – Stuart 'gently led him out', he claimed, and had a new lock put on the door. An angry Mylne then returned, was let in to take his private papers, and told Stuart that he would sue him.

Mylne still had in his possession the three books of architectural drawings he had preserved with such care, and which included some of his own. The board asked for them back, and on 28 September Mylne wrote from New River Head to say that there would be some delay before he could return 'any of the Hospital's few papers and drawings he still happened to have.' They were perfectly safe, but in confusion with his own papers because of the ungentlemanly and violent seizure of his office and everything in it by Mr Stuart. Moreover, they had to be separated from drawings he had voluntarily made at Greenwich beyond the scope of his duties, which should have been made by Stuart. He was about to go to Hampshire for some weeks and the matter must wait. In any case he had asked the First Lord of the Admiralty to review the matter. Stuart had seized his furniture, desks, books and private papers, and turned his clerk out of doors. It showed neither propriety nor justice, nor was it the behaviour of a gentleman, and Mylne could not conceive how he could ever seek to justify it.

The board, Mylne added, should know that Stuart had taken advantage of his departure to seize a store containing £2,000 worth of salvaged building materials carefully saved by Mylne for the Hospital's future use, which he had promptly – in the guise of clearing up – sold at an undervalue to local tradesmen who were apparently so pleased with the new arrangements that they rejoiced and gave entertainments – 'no wonder', Mylne wrote acidly.

The matter dragged on, and on 5 December Mylne replied to a letter the secretary had sent him on 27 November. The few books, papers and drawings that he still had were perfectly safe, and would be delivered with fidelity. If the Lords of the Admiralty chose to do nothing, and if the General Court did not reconsider their act 'whereby they condemned … the conduct of an officer unheard and uninformed in the Smallest degree of any objections to his conduct ... while he was ... pointing out the grossest neglect of, and misconduct in, its affairs... I must in justice to myself, endeavour to bring these matters to a review somewhere else; where both parties must be heard before either are censured.' The books they were asking for would thus be evidence in the case. None of them was needed for the present works; for the few scraps of Stuart's drawings, 'some of them with very illiberal notes & references' would soon show their deficiencies and absurdities for such a great undertaking. If there was any particular paper or drawing which would forward the works, he would be very happy to send it if asked.

The following April Mylne wrote to Sandwich, and his anger leaps from the page:

> When Sir Jos. Banks applied for, and Your Lordship assigned me, the Office of Clerk of the Works at Greenwich Hospital, I little dream't, I had undertaken the most troublesome, the least profitable, and the most laborious departement in this Kingdom. – For six years I laboured in that Vineyard; and to what purpose was not totally unknown to your Lordship. – I flatter myself more was done in that time, for the real improvement of the Buildings, than for 20 years before; altho' I had the total incapacity of a Surveyor, and the Peevish jealousy of a Secretary, acting as Dragchains, in these meliorations.

He summarised some of the work he had done there, and then continued:

> From thence forward every thing went wrong, nothing could please, and I have the honour to acquaint Your Lordship, that I was dismissed from Office, after presenting a formal Memorial ... without hearing, without objection, to any part of my conduct, – and even without the smallest intimation thereof; And all this, My Lord, under the Auspices of Your Lordship's Successor in Office, as soon after his appointment thereto, as the parties acting therein, could bring the matter to bear.[17]

Mylne felt a burning sense of injustice at his dismissal, which must have seemed a humiliation both in his professional and his private life. It was not in his nature to ignore a public slight, and he soon sought legal advice. As the board well knew, Stuart should have done much of the work Mylne had undertaken while in office. He was later to write that among other work, the new roof of the chapel was to his design. Moreover the Board had agreed that he would be paid for the extra work he was asked to do in the aftermath of the chapel fire. The dismissal took no account of this, and he now sued for damages. Grumbling, the board paid up, and Mylne then returned the books of drawings.[18]

Another attack on Mylne followed while he was still in dispute with Greenwich Hospital. In 1784 a city wine merchant called Josiah Dornford published a series of letters under the name *Fidelio* alleging misbehaviour and incompetence against many of the city's office holders, including John Wilkes the chamberlain, George Dance city surveyor, and Mylne who was still surveyor to the bridge. All it amounted to in his case was that no surveyor was needed once the bridge was built. This was nonsense, and when Mylne was asked by a city committee to give his version he did so at length. He said it was awkward to write about one's own actions, but he attached a list of his duties over the past twenty five years, pointing out that everything he did was done in full public view. Whether on the bridge itself – which he pointed out had been built for far less than Westminster – or its long approach roads with their new building plots and many changes to existing ownerships and premises, great quantities of work had been done by him or by the employees that he paid from his own pocket. His plans, he said, were copied and resorted to freely, his advice was always given readily and without a fee, and his house was open at all times as a public office for those purposes. He had located the owners of land that had to be purchased, negotiated the purchases, arbitrated in disputes, devised how the tolls should be collected and supervised the toll collectors and other workers, laid out new streets and ensured they were properly made up, negotiated with the trustees of the Surrey turnpike, seen to repairs when the bridge was damaged by colliding vessels, arranged for milestones and obelisks, and dealt with all manner of encroachments and problems. Those who took on public service, he thought, deserved protection from this kind of blind envy, hatred, malice and lack of charity.[19]

None of Fidelio's attacks had substance, but the matter must have been time-wasting and very annoying, especially as the letters had carried a hint of corruption against the office-holders.

While the problems at Greenwich dragged slowly towards a victory that can have brought Mylne little satisfaction, and Dornford's complaints wasted his time, another problem was in the offing. It must not be thought from this that Mylne staggered from one crisis to another. Rather, his career was marked with satisfied clients and excellent work, while personal attacks were a commonplace of the time. But successful work leaves little to write about, and it is the handful of more dramatic incidents that we must focus on.

By this time he was well known as an expert in engineering and building, and his skills were much in demand. He gave evidence for Boulton and Watt to extend the steam engine patent, and to parliamentary select committees about rivers and harbours, copper mines, and many proposed canals. He was appointed by the courts to sit as an arbitrator deciding disputes within his expertise. He must have enjoyed this work, as he never seems to have declined it, but one case in which he was a witness was to prove far more troublesome than he can have expected. It concerned the harbour of Wells in Norfolk and shows the extremes to which he was prepared to go to defend his reputation.

Wells-next-the-sea is a tiny port on the long sandy coastline of north Norfolk. It is 'next' the sea only in the sense used by continuity announcers, next but one after some interruption. In the case of Wells the intervening item is two miles of sand dunes between it and the open sea, which can only be reached by a winding, treacherous, shifting tidal channel. By 1780 this was silting up, and locals blamed land reclamation work started sixty years earlier by landowners as part of the agricultural improvements that made Norfolk famous. They had built embankments to turn eight hundred acres of salt-marshes into rich silt farmland. The marshes formerly flooded on high tides, penning back water which then drained out through the harbour creek as the tide receded. This flow of scouring water, it was claimed, had prevented the channel from silting up. Efforts were made to counteract this effect by building a sluice, but this made no difference and further land reclamation in 1758 seemed to make matters worse. Over the same period the mouth of the harbour channel had moved almost a mile to the east, forming a dog-leg entrance channel which made entry much more dangerous for sailing ships, many of which were wrecked. Old sailors came forward to attest that things had been much worse since the reclamation, and the harbour commissioners started a lawsuit for public nuisance to have the most recently reclaimed land restored to its former marshy state.

Sir John Turner, the landowner who had done the most recent work, then died and the defence of the action fell to his heirs, including Sir Martin Browne Folkes. A jury would have to decide whether the harbour commissioners were entitled to breach the embankments. Folkes asked Mylne to advise. Mylne spent a week in Norfolk in October 1780, visiting him and surveying the land with the help of local surveyors. His conclusion was firm. The silting up was nothing to do with the reclaimed land. It was part of a long-term geological process. Every time the tide came in it carried, swirling in suspension, sand from the North Sea that originated from distant river estuaries. At high tide the movement of the waters ceased for a while and the sand was dropped. As a result, each tide left the land imperceptibly higher, and this was a continuing process that had through time caused the decay of the harbour – the sea was embanking the whole coastline without human interference, and would continue to do so. The small area of reclaimed marsh would have had no significant effect on the process. He set these matters out in a report, and declared himself ready to testify to that effect.

Once the experts' reports on both sides were available a meeting was held at chambers in the Temple, to see if the experts agreed. One of the counsel for the harbour commissioners noted rather sourly that Mylne found fault with everything the commissioners had done at Wells; spoke as if he knew of a method to save the harbour; and said that if he and his opponents' surveyor could get together they would soon produce a solution. Mylne, he noted, 'seemed by his conversation to suppose himself better qualified in this Business than any other person'.[20]

No agreement was reached, and the trial – or first trial, as it would become – was held at Norwich Assizes in August 1781. Mylne took leave of absence from the New River, made a detour to view Wells harbour again and then went on to Norwich to give his evidence. The other side had called mariners and pilots but no scientific expert. The jury heard Mylne's evidence, accepted it, and found against the harbour commissioners.[21]

The commissioners then applied for a retrial, saying they had not foreseen the expert evidence and had been taken by surprise. Their application was heard by a panel of judges headed by Lord Mansfield, one of the greatest of English judges. In his judgement he spoke of Mylne 'whom everybody knows to be a man of skill and ingenuity; knows how harbours are constructed, has made Observations on the appearances of Nature, and the operation of them; and has made it his study. He shewed that the Bank cannot do any material injury, but that it proceeded from another cause.' He then held that it was nevertheless only right that one man of science ought to be answered by another, and granted a retrial, with both sides directed to serve copies of their expert reports on the other side in advance.[22]

At that stage the harbour commissioners had hoped to use John Smeaton as their expert, but they found the landowners had already booked him to provide support for Mylne. Smeaton visited Wells and surveyed the land as Mylne had done, and reached much the same conclusion, that the harbour's problems were caused by the constant deposition of tidal silt which accelerated as grasses began to colonise the new ground and trap even more silt. The same processes of nature, he thought, that had formed Wells Harbour would eventually close it so that crops would grow where ships once rode at anchor. Meanwhile the harbour commissioners found four other engineers to act for them, Grundy, Nickalls, Hogard and Hodskinson. Unlike Mylne and Smeaton they did not concern themselves with slowly acting geological processes. They simply consulted local witnesses, measured the land and reached the opposite conclusion: that the reclamation work, and especially the most recent embankment, had dramatically reduced the amount of water available to scour the channel and thus caused the problem.

The second trial took place the following year, in July 1782. This time both sides provided the jury in advance with copies of the expert reports, Mylne's as well as Smeaton's for the landowners, although they proposed to call only Smeaton to give live evidence having made a tactical decision to rely only on their best-known expert. At this trial there was a new barrister to lead the harbour team. George Hardinge was young, brash and ambitious, an old Etonian who had just entered Parliament and was undecided whether to follow his grandfather as Lord Chief Justice or his uncle as Lord Chancellor. For him this was the sort of case on which reputations were made.

Hardinge took a new and unexpected objection, arguing that Smeaton should not be allowed to give evidence. His argument was that Smeaton could not speak to the facts, he could only pass an opinion which was little more than speculation about natural processes. A jury's verdict had to be based on fact, not unverifiable speculation. The somewhat old-fashioned judge agreed, and Smeaton's evidence was excluded. Hardinge also commented unfavourably on Mylne's absence, suggesting that it was because he could not justify the opinion he had given in his written report – which the jury had seen. In the circumstances, deprived of the geological evidence, the jury unsurprisingly reached the opposite conclusion from their predecessors and found that the reclamation works had indeed damaged the harbour.

Now it was for the landowners to seek a retrial, arguing that Smeaton's evidence should have been admitted. Once again the case went before Mansfield, and he agreed. Of course Smeaton would be expressing an opinion, but it was one he was qualified to give and was based on undisputed facts that he had observed, such as the position of banks, the course of tides and the shifting of sands. Who better to explain that than a man experienced in the construction of harbours? 'I cannot believe that when the question is whether a defect arises from a natural or an artificial cause, the opinions of men of science are not to be received,' he ruled, observing that he himself had called Smeaton's evidence in the past. Lord Mansfield's judgement in the case came to be regarded as a landmark in the development of the law of expert witnesses, and *Folkes v. Chadd* is still cited in textbooks.

The stage was thus set for the third and final trial, to take place at Norwich summer assizes in 1783. This time Smeaton wanted Mylne to join him in giving evidence in person, and wrote to their solicitor:

> Be pleased to make my best compliments to my friend Mr Mylne, and tell him that I very much wish to meet him on this Sunday; or if he cannot do that; to go down together to the Assizes at Norwich. I am sure that his opinion will not only be of great weight with a jury, and therefore of great importance to the cause; but that his absence (as it seemed very strongly to operate the last time) will be considered as a dereliction of his Thesis; and this in a matter of opinion among Artists; to those who cannot (like a real Artist) see the full merit of the case, will always be considered as a drawback upon the opinion of any other, tho in direct confirmation of the fact; as I consider mine to be of Mr Mylne: I shall therefore, not only for the sake of Justice in the determination of the cause but for my own sake as an individual, wish not to stand single and unsupported against a Legion; though I think our Joint Thesis Ever so right.[23]

Mylne agreed to this, and the third trial then took place with expert witnesses called on both sides. At this trial Hardinge won again, but he did so by the use of a dubious tactic. Instead of simply challenging Smeaton and Mylne as incorrect and then calling his own evidence, he accused them of perjury – knowingly giving false evidence – and even declared he would prosecute them for it. He cannot have had any justification for this, but his apparent fervour was enough to sway a jury who had heard conflicting expert evidence on the other side and who were instinctive supporters of the fishermen and mariners of Wells whose livelihood was at risk.[24]

Whether Hardinge believed his tactic justified is doubtful. Immediately afterwards he wrote a slightly shame-faced letter to his friend Horace Walpole explaining that he had been provoked by one of the barristers on the other side who was a political opponent. As he put it, the other barrister had 'deported himself in a manner so illiberal that I could not, without injustice to the cause and my own personal honour, decline the painful task of setting a mark upon him. I was fortunate enough to carry the point, and ascribe the victory more to him than to myself. *These are odious parts of our profession.*' This is hardly the turn of phrase of counsel who is happy about his conduct.[25]

But the case was over, and Hardinge's damaging allegations were widely reported in the local press. Mylne and Smeaton were ridiculed, and Hardinge was exalted, even in verse:

> Ye men of Wells and country round forget your sorrows past,
> Let mirth and jollity abound for Justice reign at last …
> But now the day, the day's your own, your rights are now regain'd
> For Hardinge talked injustice down, and bold opponents chain'd.[26]

Smeaton had been hurt by his treatment at Hardinge's hands, and wrote peevishly to Folkes's solicitor complaining at what he described as an attempt to destroy his character, by counsel saying that he had sold himself to the side he appeared for and would give their testimony right or wrong. Despite this irritation he took the matter no further.[27] Mylne, on the other hand, was not one to ignore such an attack, and his silence over the next year proved to be an ominous one.

Mylne must have seethed, not because his valid and well-founded opinion had been rejected but because his honour and his reputation had been besmirched. Hardinge's accusation was that he was prepared to take his oath on whatever version of events would best suit the client who was paying for his services. Worse, according to Hardinge, he had gone so far as to commit the grave crime of perjury. It is a matter of record that a year later Mylne succeeded in obtaining an apology from Hardinge. This event does not seem to have been mentioned in any of the early writings about Mylne's life, while the two modern writers who have discussed the circumstances have taken it to be the result of a threat by Mylne to take legal action against Hardinge.[28]

As we know Mylne did not hesitate to seek the protection of the law where his reputation was concerned, and he was already in the throes of such a case against Greenwich Hospital. So he almost certainly took advice as to how he could vindicate his professional name from what he saw as an unprincipled and groundless attack. If so, he will have been disappointed with the answer: he was most unlikely to succeed in a lawsuit against Hardinge for the things he said in the course of presenting his client's case. Until very recently barristers were immune from such actions. Hardinge may have wildly overstepped the mark at Norwich but Mylne's chances of succeeding in court were remote.

If Mylne could not obtain redress in court, did he have any alternative? Duelling has so completely disappeared from modern society that it is easy to forget how widespread it was until well into the nineteenth century. It was usually confined to those who had the status of gentleman and was closely bound up with notions of personal honour. A gentleman who felt his honour had been attacked was entitled, and sometimes expected, to demand an apology backed by the implicit threat of deadly force. Mylne, who had claimed his right to the family coat of arms years earlier, certainly ranked as a gentleman, as did Hardinge. Mylne can be seen wearing a sword in two of the Blackfriars satires dating back to 1759, noted the purchase of another in his 1768 diary, and may well have carried pistols when travelling the country in his post-chaise (e.g. illustration 18,) Technically duelling was illegal and a survivor could be charged with murder, but duellists usually escaped prosecution. So much was it part of the social fabric that a military officer who declined a duel could be court-martialled and cashiered under the Mutiny Act for failing to redeem his honour, a provision that was only repealed after the intervention of Queen Victoria in 1843. After that time the rule was reversed, and an officer could be cashiered for giving any encouragement to the practice, but one military duel is recorded as late as 1852 and there were doubtless others.

Many duels went unreported, but Mylne was in London in 1765 when Lord Byron, ancestor of the poet, killed one Chaworth with a sword thrust in a duel at the Star & Garter Tavern, Pall Mall. Byron was put on trial but convicted only of manslaughter, and as a peer was immediately freed. Three years earlier, John Wilkes fought Lord Talbot with pistols at Bagshot. Neither shot hit home, but honour was satisfied and the duellists promptly went to an inn to share a bottle of claret. Later famous duels included one between cabinet ministers, Canning and Castlereagh with pistols on Putney Heath in 1809, and one between the Duke of Wellington and the Earl of Winchilsea with pistols at Battersea in 1829. On that occasion the Duke's second, by coincidence, was Sir Henry Hardinge, a nephew of George.[29]

Mylne was not alone in being so offended by the way he was treated in court as to contemplate a duel. In 1753 Tobias Smollett was so incensed by the treatment of his character at the hands of Alexander Hume Campbell, another barrister MP,[30] that he drafted a long letter accusing him of using virulent defamation in an attempt to destroy Smollett's character. He demanded what he called 'adequate reparation', failing which he proposed to publish the letter in the press, after which Campbell would hear from him again, but '...in another manner – as an injured gentleman…' The reference to 'an injured gentleman' is of course the clue to the remedy he had in mind. Smollett sent the draft to a friend for approval, but seems to have been advised to drop the matter.[31]

The case for duelling is set out in a book from that time, written by an army officer who had acted as a second more than once, and perhaps also as a principal. While discouraging needless duelling he argued it was sometimes unavoidable. Thus it was both just and reasonable to demand satisfaction if unjust and false imputations were made against a gentleman's honour and character. The writer summed up the prevailing view: a person whose situation and conduct in life entitled him to be called a gentleman could not exist without preserving his character absolutely unspotted by anything done or said by another gentleman. He had no choice – where circumstances demanded it, he was obliged to issue a challenge, or his reputation would be lost. Unless he did so, his character could be vilified and all his domestic peace destroyed. Moreover, duelling righted the imbalance that might otherwise exist where a rich or powerful gentleman abused the character of a poorer one: money and influence could achieve a lot but neither of them, he pointed out, would 'turn a pistol-ball, or ward off the home-thrust of a rapier.' Often it would be possible to settle the matter without a duel, and where honour permitted it the wrongdoer should of course apologise.

In dealing with the practicalities he emphasised the importance of choosing the right seconds. Seconds were necessary not only to ensure fair play but also because, where proper, they might be able to persuade the antagonists to reconcile their differences and thus avert the contest altogether or stop it after a single exchange of shots. If not they must ensure things were done properly. They must themselves be armed, in case of foul play by either side. The duellists must each have a pair of pistols of similar size, and each a sword of the same length as his opponent's, unless it was previously agreed that only pistols were to be used. Pistols should be loaded in full view to avoid any malpractice. Level ground should be used, and if the terrain prevented it then the more favoured higher position should be decided by drawing lots. The distance whether eight, ten or twelve paces, should then be measured out, and lines drawn on the ground. The parties should stand sideways on, in a fencing position to offer the smallest target to the pistol ball. Each should fire in turn, the first shot being fired by the one who claims he has been wronged. No shot must be fired until one's opponent had been asked whether he was ready and had answered that he was. The pistol must not be steadied across the other arm, and must be fired after taking no more than a moment or two to aim it. The practice of starting back to back, walking a given number of paces, then both turning and firing simultaneously should not be followed, as it might tempt the unscrupulous to fire as they turned. A surgeon should if possible be in attendance nearby.[32]

The opposite view, a classic lawyer's analysis of the case against duelling was coincidentally published in 1784, the year we are concerned with, by a barrister who also wrote treatises on suicide and what he called 'the pernicious effects of gaming'. In his view, an injury was either one for which the law provided a remedy or not. If not, to seek redress by private force would imply contempt for the law. Thus, an injured individual should look for redress in some other way than by duelling. To do otherwise would be to insult the law and to show 'a very indecent contempt' for the Parliament which makes it. Moreover, he wrote, how could any supposed law of honour force someone who might be a useful and loved member of society with many dependants, to stake his life against some worthless creature who happened technically to be his social equal but who would die unlamented? He also pointed out that duelling in law was murder, though he conceded that in practice juries, if they did not acquit altogether, might convict only of manslaughter, which then carried a maximum of one year's imprisonment, or of the trivial offence of fighting in a public place.[33]

The rules of duelling were not rigid, but the standard starting point was a challenge, often an insult. It was open to the offending party to apologise, if he could honourably do so, for example if he had been mistaken. In such a case honour was satisfied and no duel ensued. A person who felt offended could begin the process by offering an insult in return, thus provoking a reluctant or cowardly opponent to confront the issue. If the unwilling opponent then apologised, the challenger could do so in turn and the matter would end.

Seen against this background the purport of a series of letters dated in July and August 1784, a year after the final trial at Norwich, becomes only too clear. Their authors were Mylne, Hardinge, and two of Hardinge's friends, Lord Camelford and Augustine Greenland.

The first was from Mylne to Hardinge on 24 July 1784, and deserves to be set out in full.

> Sir,
> The injurious and illiberal treatment which I received, at the last Summer Assizes, at Norwich, from you, in your capacity of Counsel at the Bar, requires some reparation.
>
> A twelvemonth is almost elapsed, by the same Assize time being now come round; during which I have left you, without notice taken, at full liberty to substantiate those charges, against my character, which you threw out on the remarkable trial of the Wells-harbour cause; and, for your attempting that prosecution for perjury, which you not only hinted before you heard a word from me, as an evidence, but more than threatened, in your inflated, noisy, frothy speeches of reply.
>
> I can have but little or no expectation in this address, that a mind, of so illiberal a turn, as being capable of and actually committing such an outrage to decency, and to every quality that constitutes a Gentleman, should now feel from this remonstrance, the necessity of a proper reparation.

It is not enough for me, that the Gentlemen at the Bar felt the disgrace done them, by your audacity, in a character common to them all; and that the Judge took all the pains he could, in justice to me, on the threshold of his charge to the Jury, to do away the foul aspersions you attempted to fix on my reputation, not as an Artist, but as an honest man. – Something more is necessary; – and it must come from yourself.

The intention of this letter, is to leave it to yourself; and if you felt, as you ought, and every good man would do, on such an occasion, you will embrace this opportunity, since it has not occurred to yourself, in respect of me, as well as of Mr Smeaton.

If, hurried on by the impetus of a heated and unguarded imagination at the time, you should not now recollect, that which gave offence in so publick a manner, you can easily be informed by the Short-hand writer.

If you decline all or any reparation, you will not think it extraordinary in hearing from me again.

I am your very humble Servant,
ROBERT MYLNE.

Is this letter really no more than a polite request for the favour of an apology? In it Mylne not only explains the offence he alleges but takes the opportunity in effect to insult Hardinge as being no gentleman – a classic formula for a challenge to a duel – while the final chilling paragraph about hearing from him again removes any doubt. His objection, as he makes clear, is not to an attack on his competence as 'an Artist' or skilled professional, but on his reputation for honesty, which formed an essential part of a gentleman's honour. The letter also reminds Hardinge that the judge had viewed his allegations with sufficient disapproval to give the jury a warning, presumably that there was not the least evidence to support Hardinge's assertions, at the start of his summing up. The reference to the other gentlemen at the bar – there were at least seven barristers in the case – suggests that they too had distanced themselves from Hardinge's remarks, possibly in their speeches.

Hardinge apparently replied on 28 July, but his letter was delayed and the first reply to reach Mylne was dated 29 July and came from Lord Camelford at Petersham.

Camelford, ennobled a few months earlier, was a politician and the friend of Hardinge who had just obtained for him a seat in parliament. This made Hardinge MP for the rotten borough of Old Sarum, one of those abandoned towns that still retained its ancient rights. In the case of Old Sarum there were only three houses in the whole constituency, yet whichever landowner controlled that handful of votes – Camelford at that time – could send two members to Westminster. As Tom Paine would point out, this was the same number of MPs as the whole county of Yorkshire with its population of almost a million.

Camelford introduced himself as a particular friend of Hardinge, who had sent him Mylne's letter, with a copy of the reply by Hardinge that Mylne had not yet seen. He said he had not yet seen Hardinge to discuss the matter, but went on 'I do not hesitate under the circumstances to beg for both our sakes, that I may have half an hour's conversation with you before the matter proceeds any further.' He suggested a meeting the next day, either at his home in Oxford Street or anywhere else that suited Mylne.

Once more this is hardly the kind of letter that would result from the threat of a mere lawsuit. It must have been delivered by hand, for Mylne was able to reply to Camelford the same day. He pointed out that he had not received the reply from Hardinge that Camelford referred to, and since so much depended on its content, there would be no point in his meeting Camelford at present. He continued, in a letter that uses the language of honour, not pending litigation:

> But your Lordship may be of use, by giving him some advice, that results from a cool understanding, and sentiments of honour. He ought to give such an acknowledgement, as will satisfy the injury; nor is it unbecoming a man of pure honour to do so.

Camelford replied the following day, 30 July, saying that he had that moment received Mylne's letter, which he observed was by no means conciliatory and which he claimed he would not think of showing Hardinge. The rest of Camelford's letter shows that he had both misjudged Mylne's determination and under-estimated the extent of Hardinge's misbehaviour. He began by conceding that he knew nothing of the matter, but pointed out that whatever the provocation may have been, Hardinge had given it not in his private capacity but in the discharge of his duty towards his client and in open court. He was surprised that Mylne had not seen fit to ask for an apology at the time when the injury was fresh, but had waited a year before demanding 'in cold blood ... satisfaction in a tone that scarcely admits of explanation'. 'Now sir,' he continued, 'if this state of the matter is a true one (whatever might be the nature of your provocation so many months ago) you will do well to consider with yourself what would be your situation under consequences that might not improbably arise out of your conduct.' Conceding that warmth in the cause of his client might have taken Hardinge too far, he warned Mylne to 'reflect a little for your own sake before you suffer the violence of passion to carry you to such extremities as may be much more serious in their consequences.' He concluded by saying that while few barristers would answer out of court for their conduct in court, 'But Mr Hardinge is a gentleman and has every generous feeling that belongs to one, and would never ... hesitate to take blame upon himself where he has injured another ... though he will disdain as he ought to make any concession that can have the appearance of being extorted from him.'

What did he mean when he spoke of serious consequences? Losing a lawsuit? Or was he reminding Mylne that those who followed the code of honour must take the consequences, which might include death for the loser and prosecution for the winner. If his letter was intended to calm the situation it is unlikely to have had that effect, but before Mylne could respond to it he heard from Hardinge. Hardinge's letter was dated 29 July, and said he had just called on Camelford and learnt that, as a result of a mistake by his own footman, his reply to Mylne had not been sent as promptly as he had ordered.

There is something odd about the dates here. Camelford wrote to Mylne on 30 July saying he had just that moment received Mylne's letter, and would not dream of showing it to Hardinge. The letter he referred to was the one where Mylne told him he had not received Hardinge's first letter. Yet Hardinge wrote to Mylne on the 29th that he had just seen Camelford and learnt that his original letter had gone astray. Was one of the letters wrongly dated or was Camelford in closer contact with Hardinge than he admitted? Whatever the explanation, Hardinge now sent by hand a copy of his original letter, which read as follows:

> Sir,
> The zeal of an advocate often carries him too far into censure as well as praise; and I am not exempt from so general an imputation, tho' I trust that I am not more faulty than others in the same line with myself. But I've no scruple to say, even in answer to such a letter as I have just received from you, that if at the time of the offence which you describe yourself to have taken, you had expostulated with me in terms which left room for an apology on my part, and that I had felt that I had used improper expressions concerning you, I would have made you all the reparation which a man of honour could make for the injury you had received.
>
> The time which you have taken to resent, what I said in the Wells-cause a year ago, and the violence of temper as well as illiberal terms in which you have expressed that resentment have made it impossible for me ever to enter into the subject of your complaint more than to say that in answer to such a letter at such a time I do not think myself under the necessity of making any explanation.
> I am Sir your humble servant,
> G. HARDINGE.

Hardinge's first letter, then, had been a defiant refusal to consider the matter any further in any circumstances. His covering letter, written just a day later, but after speaking to Camelford and perhaps others, climbed down considerably from that uncompromising start. In it he declared his readiness and wish to make Mylne any degree of reparation that one gentleman could make to another, if Mylne would enable him to do it with honour by apologising to him in turn for the terms of his letter, or by rewording his demand so that Hardinge could answer it by explanation or apologies. He said he distinguished the substance of Mylne's claim from its manner. He asked Mylne to consider this for two or three hours, after which he would 'send a particular friend of mine to receive your answer as well as to confer with you if you should think proper.'

The 'particular friend' – or potential second – was one Augustine Greenland, an attorney and a protégé of Hardinge's uncle Lord Camden, the former Lord Chancellor. When they met Mylne made at least two things clear. The first was the precise nature of his complaint: he had obtained a copy of the court shorthand writer's note from Norwich of what Hardinge had said, so was able to show him verbatim the enormity of Hardinge's 'odious parts'. The second was his reason for delaying a year: if he had complained any sooner Hardinge would have obfuscated with talk of a pending prosecution for perjury. It is tempting to think that a third topic was also discussed: removing any doubt what the consequences would be in the absence of a full and public apology.

After that visit, Mylne made time to reply to Camelford, whose long letter of the 30th, he observed, 'proves how little your Lordship is informed of the magnitude of Mr Hardinge's atrocious behaviour towards me, in 1782 as well as 1783; not a hasty and spontaneous eruption in the zealous efforts of a pleader; but a cool and predetermined attack on my character, on purpose to serve his side of the cause.' Of this, he said, he had convinced Mr Greenland. If such was the general practice of the Bar it was a great disgrace. But in any event, general practice was no excuse: bright talents disdained such prostitution, and Hardinge's prospects led him to a nobler conduct. Camelford would do well to throw a little ballast aboard Hardinge's vessel, and it would sail the better for it once this little storm was weathered. Camelford was quite correct to say the offence had been given 'in open court'. That made it worse, not better, and unlike a private injury it could not be put right by a private acknowledgement. As to the year's delay, it was to give Hardinge 'all the routine of Terms and Vacations to do that which he assured the Court he proposed to perform', i.e. to start a prosecution against Mylne and Smeaton. To have complained sooner would just have given him an excuse. Mylne concluded by pointing out how Camelford's understanding of the facts, as gleaned from Hardinge, had proved untrue, and how Hardinge had thus once again wronged him, whether by want of memory or otherwise.

On the same day Mylne wrote to Hardinge, to acknowledge receipt of his two letters and Greenland's visit. Then he came to the point. He had shown Greenland the independent note of what Hardinge had said, written down in court while he was saying it, to show that he had just ground for the complaint he had made and the reparation he expected. 'I have only to say it must be full and explicit, as well to the Assizes of 1782 when I was not present to take such matter up; as of that of 1783, at which your preceding behaviour was the principal cause of my attending.' As to any apology from Mylne for what he had said in his original letter to Hardinge, that could only begin to be thought of after Hardinge had completely restored his character.

The same day came a letter from Greenland: he had been in touch with Hardinge. The apology would be forthcoming and Greenland would personally deliver it into Mylne's hands. This did happen but, as Mylne later put it, this first version of the apology 'did not appear sufficiently explicit and satisfactory to him, and indeed so much qualified and guarded that he returned it as inadequate.' Finally, on 4 August came the letter Mylne wanted:

> To Robert Mylne Esq.
> In answer to your requisition which you have made by Mr Greenland, I assure you that I am extremely concerned for the Offence which I gave you at the Summer Assizes for Norwich 1782 and 1783; when, speaking as an advocate in the Wells Cause I charged you with perjury and menaced you with prosecution for it.
>
> I lament sincerely, that my Zeal, as an advocate, in both hearings, betrayed me into such reflections, which could not be Justified by your Conduct or Character, and for which you had a sufficient ground of claim to every degree of reparation, that a Man of honor can make.
>
> I mean to convey the Idea of such a reparation, in the utmost extent of it, by the Words that I have used, and am Sir,
> Your most humble servant,
> G. Hardinge.

Hardinge also sent a letter to Smeaton in similar terms. Mylne in turn did the honourable thing, with a reply to Hardinge that acknowledged his apology and saying that he in his turn was now 'sorry for the disagreeable manner and unhandsome expressions' used in his first letter. So it was that within just seven days Mylne had obtained precisely the apology to which he felt entitled for himself and for John Smeaton. Whether recourse to law would have obtained it in as many months, if at all, seems very doubtful.

Up to this point the exchange had all been private, but Mylne had made it clear that the apology must be as public as the offence. This he achieved by having the entire series of letters printed, including Camelford's, and sent to 'the Judges, Jury, Counsel, Ingineers and other evidence who attended the several trials at Norwich on the Wells-Harbour cause; and to such persons as are or may have been any ways interested or conversant in the principal business thereof,' with a footnote to point out that the jury's decision had been dependent solely on the character and credibility of engineers for honour, integrity and soundness of understanding. Copies of this printed bundle of letters survive in at least three archives.[34]

One entry in Mylne's diary may give a clue to his feelings about this affair. The final version of Hardinge's apology came on 4 August, and the following week Mylne made a note of what for him was an unusual piece of extravagance – the purchase of a gold top for his walking cane, unique in all the years he kept his diary. Was it his present to himself, for having weathered the storm to redeem his reputation even at the risk of being ready to endanger his life?

Predictably, there is very little else in Mylne's diary to illuminate these events. The fact that he visited Norfolk to survey the harbour and later give evidence is recorded in the usual terse entries, as is his payment to Blanchard the shorthand writer for supplying him with a note of Hardinge's remarks and to his friend Woodfall the printer for printing copies of the whole correspondence for circulation. But of his anger and frustration during the intervening year there is not, of course, a single word.

There was a sequel to this. Three years later Mr Blanchard published a book, The Complete Instructor of Shorthand, quite expensively priced at one guinea. Like many books of the time it begins with a short list of the subscribers who had signed up for copies in advance and made the publication possible. It is no surprise to see Robert Mylne's name there, but also listed is the name of George Hardinge.

As for Smeaton, he was so pleased to receive his copy of the apology that he wrote back to thank Hardinge for it. He was one of the first users of Watt's copying press, so there is a copy of the letter among his papers. He explained that ill health at the time, which made him consider giving up his professional work, had prevented him from seeking an apology as Mylne had done, and continued:

> Yet like him I could not help feeling and thinking hardly of a Man who had closed my professional Appearance in a Court of Judicature in the Stile of a Witness, who was about to give a premeditated false Testimony, after I had firstly been under the handling of almost all the great Council at the Bar; from De Grey, through the present Lord Chancellor, Lord Ashburton &c down to Mr Erskine: no one of whom had ever thought of treating me in terms of disrespect: however the Sharpness of your Modes of Expression in a great Measure lost its force upon me, by the full Confidence, that no Man who knew me, would believe a Syllable of the Insinuation but that it would naturally reflect a discredit upon him that used it.[35]

Lord Camelford's well-meaning attempts to head off a duel turned out to be sadly prophetic. Twenty years after these events his only son and heir, a wealthy and rather wild young naval officer, provoked and lost a completely unnecessary duel at Kensington against his friend Captain Best, said to be the best shot in England, and expired painfully three days later from the effects of a pistol ball that pierced his lung and lodged in his spine. The Annual Register, while praising his fine character, called him a victim of his own impetuosity and deplored the modern system of manners that encouraged such a waste of life.[36]

As for Hardinge, Mylne's circulation of the entire correspondence, which revealed his swift collapse after initial defiance, cannot have raised Hardinge's standing in the law. Perhaps it was one more indication of what people already knew. Six years earlier, one of his friends had written to his brother the British ambassador at Madrid, that Hardinge was losing out to others with fewer advantages because of his wild conduct.[37] Although well connected it is clear that he irritated many he met. He bought a cottage by the Thames at Twickenham, close to Strawberry Hill, thus becoming a neighbour of Horace Walpole to whom he wrote begging for friendship. Walpole had mixed feelings about him, and wrote to tell others how he snubbed Hardinge's invitations and requests, calling him an 'out-pensioner of Bedlam' – a madman. He never achieved the kind of legal advancement he had hoped for, rising no higher than a middling judicial post in rural Wales: small success for one who started with his advantages. After his death Farington noted that high promise had ended in disappointment because of his irregularity of conduct, but maddeningly does not hint what form it took.[38]

Not long after the Wells affair Mylne was once more consulted on the subject of harbours, but in very different circumstances. In 1786 the British Fisheries Society was founded by a group of Scots including the Duke of Argyll, to improve the quality of life in the highlands by providing the infrastructure for a fishing industry. Some landowners were turning their estates into sheep walks, where a single shepherd could care for vast numbers of sheep in glens where there had once been whole villages of crofters. This encouraged emigration and consequent depopulation that the founders saw as a great social evil, for Britain as well as Scotland. The lack of roads made it almost impossible for the remaining highlanders to sell any surplus they could grow on their unimproved land and they eked out a marginal existence, lacking the money to trade so that they had to make every necessity for themselves, down to clothing and shoes. Although the surrounding waters teemed with fish, there was nothing to help prospective fishermen. There were no harbours where fish could be landed for sale and nowhere to buy the necessary supplies, not just hooks, lines and nets but also barrels and the masses of salt needed to cure herrings. Salt was a particular problem as it was taxed and supplies were therefore controlled by the Customs. Any salt used to cure herrings was duty free but its supply was subject to a network of regulations that made it almost unobtainable to fishermen who did not have a local custom-house. Any such salt had to be taken to a custom-house and weighed by officials. Then the user had to take all his cured fish back for inspection during the same tax year, and swear an oath that all the exempt salt had been used in its curing. Even then a lesser rate of duty was still payable if any of the fish was to be sold, and severe penalties were imposed on anyone dealing in uncertified fish. The Society tried to solve these problems by founding new towns at suitable locations, and among the first were Ullapool on the mainland and Tobermory on the island of Mull. Here families would be encouraged to settle and build houses, traders would be allowed to build stores and sell necessaries at fair prices and schooling would be provided.

In 1788 Mylne prepared a long report on the harbours, piers and houses that would be needed. In the next two years he was asked to provide some designs and did so, including plans and estimates for works at Ullapool and designs for an inn and a custom-house at Tobermory. All these he provided free of charge, as he was ready to do when it was for the public good. In accordance with the society's wishes the designs are for much plainer buildings than might be expected from Mylne. This is explained by one of the society's main supporters who wrote 'in the beginning all magnificence ought to be carefully avoided ... until the company's settlers have amassed as many crowns from herrings as Birmingham has from hardware.' The founders knew how easy it would be to dissipate their limited resources and wanted to give the fledgling industry its best chance for success by prudent housekeeping and simple unadorned buildings.

Sadly a movement of the herring shoals away from the west coast meant that the new industry languished on that side of the country, and Tobermory and Ullapool were already in decline by the end of the century although the Society's other main settlement, Pultneytown at Wick on the north east coast thrived and became Scotland's main herring port.[39]

Meanwhile family life had been going on in the background. A few weeks after Mylne's dismissal from Greenwich his infant son Tom had died of smallpox and for a time there were no more, but then came two more girls, Charlotte in 1785 and Leonora in 1788. Apart from these domestic matters the 1780s were years of struggle and anguish for Mylne. The rioters' damage at Blackfriars was not something to take personally, and it was soon put right, but Greenwich and Wells were different matters. Despite his legal victory against Greenwich Hospital, Mylne was deeply hurt at the way the board had treated him, and a money settlement did nothing to salve the hurt. Almost three years after the event, when his successor, William Newton, wrote to him with an enquiry about some buildings at Greenwich, he replied that the letter had 'wakened me, once more, to a sense of all the wrongs done me by Greenwich Hospital...' He asked Newton to tell the directors that 'when they are emancipated from the principles which has dishonoured the Hospital so much ... they will feel a sense of gratitude for the many & essential services I did that institution, far beyond any ever done in my department...'[40]

The trouble at Wells intervened after Mylne's dismissal from Greenwich but before his action against the directors was settled, and this can only have intensified his determination to vindicate his reputation. The apology that he forced so dramatically from Hardinge, however necessary, can only have gone some way to cure the problem, for he must have known that it is sufficient for an allegation to be made for some people to believe it: some people always believe the worst, and no retraction or lack of evidence will persuade them otherwise. It would be interesting to know how much Hardinge knew of the trouble at Greenwich, and whether it tempted him to attack a witness whom he saw as already wounded. The gold top Mylne bought for his stick suggests a mild satisfaction at the outcome, but only serves to emphasise how unpleasant the whole episode had been. As the 1780s drew to a close he must have hoped that the efforts he had made to regain an unblemished reputation would bring an end to his troubles, and that happier times lay ahead.

10

'... sufficiently disagreeable sensations...'

By 1790 Robert and Mary had a houseful of children. The four oldest girls, Maria, Emilia, Harriet and Caroline were spaced a year apart, from eighteen down to fifteen. Then there was a four-year gap – marked by at least two miscarriages – before Robert and William who were eleven and nine. Then there was another four-year gap before Charlotte, aged five, and finally Leonora who was two.

Since the loss of Mylne's work at Greenwich and with it their occasional home, the family lived all year round at the Water House at New River Head. In most respects this was an excellent family home. Mylne had improved and enlarged the original 1613 building, encasing the old parts in new brickwork and later altering part of the roof at his own cost to provide extra rooms for his brood (see illustration 2).[1]

The finest room in the house, known as the court or oak room, was added by John Grene, a wealthy shareholder who was also clerk to the New River Company in 1693. It still exists, preserved within the 1920 building that now occupies the site. It was originally upstairs at the front of the house, and had windows on three sides, looking east, south and west over a cityscape dominated by St Paul's and Wren's forest of churches, all on the far side of a wide expanse of fields that separated New River Head from the village of Clerkenwell. The walls of this room were panelled in oak adorned with magnificent carving in the style of Grinling Gibbons on themes of fishing and hunting, with the royal arms over the fire. The elaborate plaster ceiling is covered with relief panels showing outdoor life with a watery emphasis of herons, swans, dolphins, and water gods, heraldic devices and an oval central panel with a painting of William III. It was a very grand room and the Mylnes must have loved it.

This gem of a house had a hidden flaw. In keeping with its original function the basement was still a cistern of water, fed from the adjacent round pond into which the New River constantly flowed. Most of London's water supply flowed through this cistern, which was connected to stopcocks that fed the vast network of wooden mains, each of them turned on for just a few hours twice a week, on a fixed rota. Boswell, after dining there, said that Mylne had shown him the wonders of the deep. This meant that, despite its elevated position on Islington Hill, it must have been as damp as if built on a jetty over the Thames: with hindsight not a healthy situation for children or adults.[2]

Mylne, whose office was on the ground floor just over the cistern, must have had a strong constitution to keep up his constant round of work and travel, but by 1788 there were signs of deterioration. In March 1788 instead of attending the New River board meeting as usual, he wrote a letter to the board proposing improvements to the system and explaining that he had been confined to his bed since Christmas with a rheumatic complaint. In early April he attended a meeting, but the minutes note he was not yet recovered and in July he was given a month's leave, later extended for a further week, 'for the recovery of his health'.[3]

His diary makes no mention of this – there are very few entries in those months, but that is an occasional feature of most diaries including his. What it does show is that he then travelled to Buxton to take the waters in August. He wrote to Matthew Boulton before that visit, encouraging him to come to Buxton as well and make up a party if he could spare the time. In that letter Mylne called himself 'a poor stiff leg'd creature: capable only to sit and talk'. He said he would have one of the boys with him, presumably Robert, but that his wife 'will not be prevailed upon to leave the Chickens she has hatch'd' – Leonora was only a few months old, and Charlotte still a toddler. He also had to stop off to give expert evidence at Carlisle Assizes about a bridge of Smeaton's that had been swept away at Hexham.

On the way he met his sister Elizabeth by arrangement at Newcastle. Her husband Robert Selby had died a few months earlier, and he was fond of Elizabeth and her daughters, for whom he always took presents on his journeys north. One, Jane, was now married to Captain Thompson RN, but the other daughter Elizabeth was very ill with consumption. A few days later, at Buxton, Mylne wrote to his sister having heard the child had died. Elizabeth had now lost a husband and a daughter in four months, and he suggested she might like to stay at Powderhall for a time while her affairs were sorted out – he was the executor of both the husband's and the daughter's estates.[4]

As to family life, most of the details have long disappeared. Mary Mylne kept housekeeping books, renewed every year like the diaries and sometimes bought on the same occasion. They have not survived, and it is only where Robert paid a bill and noted it in his diary that there is a record. From this incomplete source a picture emerges of the kind of family life that might be expected in a prosperous educated household of the time. There were servants, including a footman, a cook, a housemaid or two, and often a coachman. Study was not neglected. Robert junior – known as Robin within the family – had been sent away to boarding school when he was eight, William following him two years later at the same age. For the girls there was a succession of tutors, to teach them French and

Italian among other accomplishments. There was time for music, including the major purchases of a piano and a harp. This is not surprising; Mary's sister Anne Home Hunter, who had married Dr John Hunter the anatomist, was a poet and a friend of Haydn who set some of her verses to music, including *My mother bids me bind my hair.*

Into this busy household the first sadness came in March 1790 with the news that William had died in Dublin. Three years earlier the corporation had presented him with a magnificent silver dish to praise his work for them, and he had at once sent it to his namesake, Robert's younger son, as though he knew how ill he was. After his death, Robert arranged for a wall tablet to be placed in St Catherine's church in Dublin where he was buried. It reads:

> To the Memory of WILLIAM MYLNE, Architect and Engineer From Edinburgh: Who died aged 56 March 1790 and whose Remains are laid in the Church Yard adjoining.
>
> This Tablet was placed by his Brother ROBERT MYLNE of London to inform Posterity of the uncommon Zeal, Integrity and Skill with which he formed enlarged and established, on a perfect System, THE WATER WORKS OF DUBLIN.

Although quite specific about his skills and his origins, it is notably secular, with no hint of eternal rest or other religious belief. It confirms the impression that humankind and human posterity were what mattered to Robert. As became clear from later monuments he was to design, both for family and strangers, he set great store by monuments that sent informative messages to future generations.

A year later in 1791 that line of the family ended completely, with the news that William's illegitimate son Willy, a Lieutenant in the East India Company's artillery, had died in Bengal. He was still in his early twenties and the older girls must have wept as they remembered all the excitement of fitting him out for his journey to the east eight years earlier. By his will, made earlier that year, he divided what can only have been a small inheritance equally, half to Robert's sister Lady Anne Gordon who had cared for him in childhood when his father fled Edinburgh, and half to Robert's wife Mary for life, and then to Robert's daughters. His will was dated in March 1791, so he must have known by then that his father was already dead, as presumably was his mother – a 'poor unhappy woman' according to Lady Anne, who had forbidden any contact between her and the child on the basis that it could only be bad for his credit and morals – for neither are mentioned in the will.[5]

Despite the tensions that must have arisen from the collapse of the North Bridge, and William's flight to America at a time when his fees had not been agreed and when Robert was out of pocket as one of his sureties, the brothers had remained friends, as shown by at least one joint holiday, and Robert must have felt the loss at the extinction of that branch of his family. His sisters were variously ageing or dead, he had no other nephews, and only his two sons remained to carry forward the name of Mylne.

The next blow came in 1794. It was a busy year for Mylne who was increasingly in demand for new canal schemes, retained to advise both promoters and opponents, and regularly giving evidence before Parliamentary committees. He also found time for one family matter. In the course of his frequent journeys along the winding course of the New River he had become fond of the Hertfordshire countryside around Amwell, one of its original springs. He began by buying a cottage there for his younger son William, and then went on to buy more land in the parish of Great Amwell as it became available, some of it adjoining the New River, until he had an estate of about 240 acres there.[6] Here he planned to build a country retreat, and on 4 August he set out the foundations so that construction work could begin. The rest of the month was crowded. There was a new canal proposed for London, a week at Gloucester in connection with its new ship canal, as well as a meeting about a canal proposed for Bristol. Then it was back to London for more meetings and surveys of property, a trip along the New River to check the progress of works, then back to London again. There he had a meeting with the Stationers' Company and other work before setting out for Scotland as he liked to do in late summer. He can have had little time to spare for his large family in those weeks. Near the start of the journey there was a sombre duty to perform at Amwell – marking out the site of a grave for a friend, Mr Small, who Mylne and some medical friends had been supporting in his old age and who had just died. Then he continued north, taking his older son with him. Robert was fifteen and had finished his schooling at Charterhouse, and his father probably wanted to show him the house he would inherit at Powderhall and the family properties in Edinburgh. It is clear from the diaries that although he bought investments for the younger boy William, such as future annuities and the cottage at Amwell, and similar annuities for his daughters, there is nothing of that kind for the older boy. It was unnecessary – Robert would be his heir and become the head of the family, and it went without saying that he would inherit the Scottish properties as well as the bulk of the English estate. Their journey north took five days as usual, and while in Edinburgh Mylne entered the boy for some further tuition in mathematics, Greek and Latin.

It was while they were there on 25 September that the news came that Robert's eldest daughter, Maria, had died five days earlier, at the age of twenty-two, so they hurried back to London. The diary does not mention the cause of death, nor any unusual medical expenses, but London was not a healthy city. Maria had some artistic talents, as her miniature of Robert, made into an engraving, shows (see illustration 4). The same month his sister Elizabeth died, so that his only surviving sibling was Anne, Lady Gordon. She bore him little affection, or so her letters suggest, and was herself newly widowed as Sir John had died that year.

Mylne was still friendly with Matthew Boulton in Birmingham, and after Maria's death he ordered three gold lockets, all to the same design, from Boulton's catalogue, adding 'when they are ready, the hair shall be sent'. Presumably he meant locks of Maria's hair for the three older girls Emilia, Harriet and Caroline, who were all over eighteen by then. Eight months later he had to write again, to remind Boulton that the lockets had come without a bill – 'let me know how much I am indebted and oblidge.' He always paid his debts.[7]

The deaths meant a sudden diminution in the family, while at the same time the children were growing up. One of the last things he had done before leaving for Edinburgh was to apprentice Robert to the architect Henry Holland who was completing the transformation of the fields near Sloane Square into a fashionable suburb when he could spare the time from other projects. The apprenticeship is recorded in the register of the Stationers' Company, to which Holland belonged, as architects still had no professional body beyond the informal Architects' Club. Evidently Mylne intended Robert to be an architect, but almost nothing is known about the boy or his abilities. It did not follow that he would spend the whole six years of apprenticeship working for Holland, but presumably he spent some time with him.

The Architects' Club had been founded in 1791 with Mylne as one of the original members, like the Civil Engineers, and met monthly at the Thatched House tavern in St James's, for dinner at five with the bill sent up at eight. Some members seem to have wanted a merely social grouping, but Mylne was one of those who thought that a professional body should regulate its members' conduct and decide what kinds of behaviour would be damaging to the honour of the profession. There were sometimes personality problems, and Mylne intervened to try and resolve one that arose between Holland and George Dance, and clearly felt that the club could serve a useful function in setting and maintaining standards. By 1797 there were thirteen members, each paying seven guineas a year. They held an annual dinner, and the diarist Farington's table plan for 1799 shows Mylne sitting next to his old friend Caleb Whitefoord, presumably his guest.[8]

Despite his support for the club, Mylne was not one to compromise his integrity. In December 1790 the audience at a prize-giving at Somerset House, built by Sir William Chambers, 'was alarmed by a very loud crash underneath them, which was followed by another before they had the wisdom to consider themselves in danger. It was found by inspection that two of the beams that supported the rooms allotted to the Royal Society had broken.'[9] A committee of well-known architects was appointed to prepare a report for the Treasury and declined to find any fault on Chambers' part, except for Mylne who submitted his own dissenting report. Chambers use of fir instead of oak for the principal timbers of the floor was not judicious, he thought, while the underlying structure he had built to support the floor of a room 53ft by 43ft was not strong enough and fell far short of the merit of the rest of his work.[10] If Mylne was right about this, why did all the others disagree? If he was wrong, why did the beams break? We cannot know, but it does confirm that he was not one who just follows the crowd.

Although he continued architectural work for many clients, including the continuing improvements at Inveraray for the Duke of Argyll, Mylne's practice at this time had an increasing amount of water-related work. He designed a waterworks for Norwich, surveyed the shoals of the River Severn for planned improvements and prepared reports on improving the navigation of the upper reaches of the Thames, where millers impounded water in a way that often prevented navigation for weeks on end, sometimes leaving London short of food. It was the most frenzied era for canal building, and he advised on many. The only one in which he acted as chief engineer was the planned Gloucester and Berkeley ship canal, later completed as the Gloucester and Sharpness, which planned to turn Gloucester into a major inland port. Although much of the Severn carried heavy commercial traffic, one tidal section south of Gloucester suffered from almost insuperable navigational difficulties. It was a tortuously winding stretch, plagued with sandbanks and subject to strong and unhelpful prevailing winds. Added to this, its tidal fall was said to be the second largest in the world, and in practice even small vessels could only use it on a few days each month. This prevented Gloucester from developing as it might otherwise have done, and a solution was potentially very profitable. Mylne subscribed for shares in this ambitious project and was asked to be its engineer. Unfortunately the management had no idea of prudent financial control, inducing further investment by paying dividends even before the canal was built, which could only have been paid from the investors' own capital, something Mylne complained of. One recent analysis of the canal's problems suggests that there may have been corruption on the part of the resident engineer, and there were clearly opposing factions within the company.

What was planned, and eventually built, was a ship canal larger than any other in Britain and of great utility, but Mylne had fallen out with its committee of proprietors long before its completion. He surveyed the route in 1793, and his plan was the basis on which the necessary Act of Parliament was obtained the following year. Sadly, his somewhat precise estimate, £121,329 10*s* 4*d*, turned out to be far too low because of labour shortages, rising prices and some disastrous storms. Worse still, his commitments prevented his frequent attendance, so that his role declined from that of salaried engineer in 1794 to a consultant on a daily allowance in 1797 and an effective dismissal not long after that. Work on the canal had to stop for lack of funds in 1799, and it was not until 1820, years after Mylne's death, that it was completed to the stage of being able to take commercial traffic.[11]

On the personal front, one of the investments he had made with his final payment for Blackfriars Bridge was to buy the leases of some building plots along Bridge Street, which he had laid out to link the northern end of the bridge to

Fleet Street, over the course of the River Fleet. On one of these, on the corner of Little Bridge Street, he built a substantial house, and in March 1789 he wrote to Matthew Boulton asking for help. He wanted to obtain a licence for it as a coffee house and hotel, but this would need some support. Could Boulton find 'half a score of Gentleman of known Rank' from each of the great manufacturing towns of Birmingham and Manchester to sign petitions supporting him?[12] Presumably this was done, and what was variously known as the York Tavern, York Hotel or York Coffee-house came into being. All through 1790 the work continued, with a wine-cellar, much bell-hanging so that rooms could ring for service, some of the new plate glass and an expensive sign. By the end of the year the work was finished, and Mylne's diary notes that he gave 'an entertainment' there for his wife and children.[13]

It was around this time that his lease in Arundel Street came to an end. It had been bringing him an extra income for many years once he no longer needed to live within sight of the bridge, and it may have been the prospective loss of that that had encouraged him to start the new venture. Mylne did not run the York, but let it to a Mr Varley who shared the cost of obtaining the licence and paid an annual rent. Varley kept a good cellar, and his rent was often partly offset by Mylne's purchases of brandy, sherry, port and other wines. It was a smart establishment, and Mylne was able to use its rooms for some of his work, such as the Lancaster Canal arbitration in November 1792. A vivid description of the York can be found in *Roach's London Pocket Pilot* for 1793:

> A spacious building with noble apartments; and the coffee-room itself the most elegant perhaps in England; the proprietor not above his business and all the attendants active and obliging. This house is famous for giblet soup of the finest quality, and the bar being ornamented by one of the mildest, modestest, prettiest, best dressed, and most obliging bar-maids in the world – her manners and address easy; and unaffectedly insensible, of her conciliations, she secures the attachment of every customer.[14]

With a recommendation like that, who would not wish to look in? If it raises the suspicion that Varley wrote his own entry, it can only be said that other premises are described in such terms that their owners might have considered legal action. The York survived until 1863 when it stood in the way of railway improvements and was demolished to make way for the long-vanished Ludgate Hill Railway Station.[15] Thus, like so many of Mylne's buildings, almost nothing is known of its appearance or layout, though it is said to have had his crest and initials on the outside, and some classical medallions on the interior walls. It was certainly well within his powers to design the 'most elegant' coffee-room in England as can be seen from his designs for Almack's or the interiors at Inveraray Castle, and it provides another example of his versatility.

Meanwhile work continued at Amwell. A house built by an architect for himself is likely to be designed with care, and its design should tell us something about Mylne. He was sixty when work began, and despite the illness five years earlier still had almost twenty years ahead of him. He had no financial constraints and was well aware that building could be a good investment. At Amwell there were no planning laws to limit his design. The house was in a secluded location on ground that ran down to the New River, and he could build whatever appealed to his own tastes and needs. Although the house is still there, it has been greatly altered over the years, making the original design almost impossible to see. Fortunately some of the drawings, by his assistant, have survived and from those we can see its layout and general appearance (see illustrations opposite).[16]

It was elegant but plain, a brick house with a tiled roof and very little exterior ornament. The only decoration to the facade was a small pediment over the front door and a dentillated brick cornice under the parapet. It was of course designed to be heated by open fires and lit by oil lamps and candles. Probably for this reason Mylne set the main chimney-stack in the centre of the structure, an arrangement that he favoured. It meant that the interior would benefit from the warmth of the flues, and left the outer walls unencumbered except at the corners which each had one further chimney, so that all the principal rooms had windows on two sides for best views and natural light. The corner chimneys were its most distinctive feature. The ground floor was raised 2ft above ground level to improve the outlook and to allow generous amounts of light to reach the basement windows. Two single storey wings increased the area of the ground floor, one extending at each side. These were set back, so as not to reduce the light reaching the side windows, and attached to the main house only at its back corners. One of these wings was to be Mylne's library, the other a kitchen.

Approached by a short flight of steps, the front door led straight into a large hall-cum-drawing room. Beyond it a passage gave access to the dining room, breakfast room, library, staircase, kitchen and back door. Beyond the back door was an enclosed courtyard with a well at its centre, flanked by the library and kitchen wings and closed at the back by a semi-circular range of single storey buildings containing a dairy, bakehouse and store-rooms.

The main part of the house was large enough to give five principal bedrooms on the first floor. The two side bedrooms at the front had their fireplaces in the corner of the room, between the windows and opposite the bed, an aesthetically pleasing arrangement that made efficient use of the available space. Above, on the top floor, were sizeable attics lit by dormer windows.

The name he chose for the house was *The Grove*, and he planted trees around as building progressed, but the name had a classical as well as a literal sense: it was to be a place of studious retreat in his last years. Having at last published his map of Sicily in 1788 he was still anxious to complete his book on Sicilian Antiquities, and had written to Sir William Hamilton at Naples complaining how

Above and below: Plans for The Grove, Great Amwell. By 1794 Mylne had bought 240 acres of land adjoining the New River at Great Amwell near Ware. Here he built The Grove, a modest but comfortable country retreat where the family moved each summer. (Hertfordshire Archives and Local Studies)

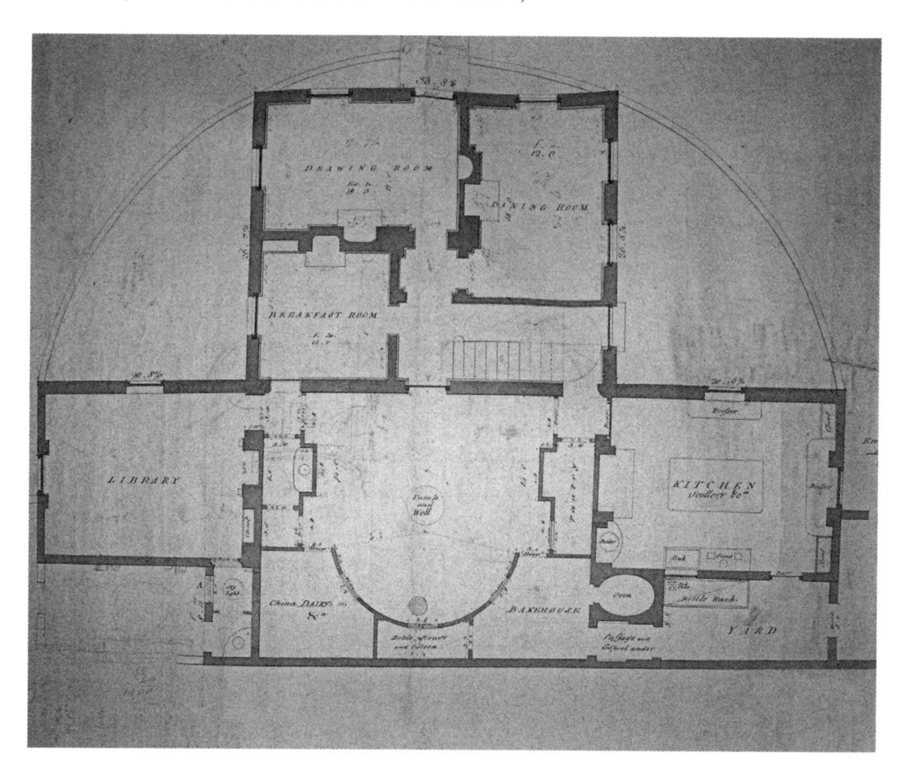

it was being delayed by 'infirmities and idleness' that crept in on his time.[17] Much of his professional work now involved writing reports, and the library suited such purposes, with a skylight and its own side door to the garden giving him some independence from the bustle of a large household. The library and the matching kitchen in the opposite wing were the two largest rooms in the house, perhaps suggesting a comfortable and equal division of roles within the family.

The whole house confirms Mylne as a designer whose sense of the aesthetic was always tied to the practicalities of efficient design. Whereas Blackfriars Bridge or the state rooms at Inveraray had called for magnificence, his own house did not, and there was nothing ostentatious about it. It was a home built for the comfort and ease of its occupants, and its simple exterior has only the decorative cornice and a parapet curving up to meet the corner chimneys to hint that it was anything out of the ordinary.

Designing it was one thing but the practical aspects of building it were another matter. Here Mylne, for all his famous irascibility, was no more immune than anyone else from the problems that beset the absentee employer of labour. It did not merit a full time clerk of works and Mylne could visit only when he could spare the time: the journey from New River Head in his chaise rarely took less than three hours each way, although his work along the New River sometimes brought him closer. Worse, he could only employ country workmen unaccustomed to deadlines or the pace of city life, men who undoubtedly felt that they knew how to pace themselves and how the job ought to be done. His diary accordingly makes occasional exasperated note of the resulting problems of drunkenness, absenteeism and incompetence, with some men walking off and others sent packing when they failed to mend their ways.

In 1796, as the new house neared completion the second daughter, Emilia, fell ill. Maria's death two years earlier must have made this a worrying development. Eminent physicians were consulted, Doctors Fordyce and Pitcairn, but her condition remained the same. As the months went by it did not seem too serious, and when Mylne wrote to his friend James Watt about a steam engine in February 1797 he reported 'We are all pretty well, coughs & colds excepted.'[18] The third surviving daughter, Caroline, then became betrothed to Colonel Duncan of the East India Company, a cousin of her mother's, and the wedding took place that June at the parish church of St James Clerkenwell, just a few minutes' drive by carriage from New River Head. It was understandably a great family occasion and even Mylne's prosaic diary conveys some of the excitement and bustle of new outfits and hats for the girls, and the piano he bought for the couple.

It was just after this that the new house at Amwell was at last ready for occupation, and on 8 July 1797 Mylne noted that he had moved his wife and all the family there. Predictably, there is no description of how they moved in and spent their first days there. Mylne stayed there as well until 11 July, but then returned to London to work as usual. The next entry he made is all the more shocking for being so completely unexpected: '13 July. Mrs Mylne died at Amwell, at 7. Morning. Went to Amwell. Sent family to town.'

He always wrote of his wife as Mrs Mylne, following the formality of the period. Nothing before this in the diary suggests any problem with Mary's health. He stayed at Amwell, presumably making funeral arrangements, then returned to London and his family the next day. The funeral took place at Amwell a week later, and an announcement in the *Gentleman's Magazine* explained the cause of death as a complaint in her lungs. Mary was just forty-nine, and they had been married for twenty-seven years.[19]

Mylne occasionally confided to letters the things he never noted in his diary, and fortunately one to James Watt survives from that time. It seems that Mrs Watt had sent a chatty note that had to be rapidly followed by condolences as the news of Mary Mylne's death reached her.

London, Aug 26 1797.

My Dear and Very Good Sir,

I am hardly able to return you a suitable answer, to your kind & friendly letter of the 21st – I felt for you & Mrs Watt not having known, what had happened; while she was communicating her pleasant and agreeable sentiments &c, and to still acquire the knowledge of it, thereafter.

We have borne up, as well as we could, but, in fact, it has shook our little system, to the Centre.

The marriage of my Daughter Caroline, tho' perfectly satisfactory, yet the parting for life; and the fitting out, as well as parting for life also, with Robert, on going to Gibraltar; had added to all our sufficiently disagreeable sensations. – And now, there remains the long illness of Emily, to be got the better of, and the patience necessary for such attentions as it requires. – Mrs Watt and you have been uncommonly kind in your obliging tenders of trying change of air &c. I wish that could do. But I doubt, whether she has strength to support that, which would be fatigue to her in so great a transit; while she might benefit, in respect to air &c.

She has lately acquired a great deal of strength slowly, but she has a cough and no vigour. – I am going to try an experiment, of driving a little way at a time in an open 4 wheel low chaise. From which we can better judge, of great Distances and rougher Roads.

From the same reason, we have declined going to Lady Thompson at Southampton. – We are tolerably now in Spirits; finding that real grief is the best cure for itself, in peace and retirement. – Some friends who can bear serious moments, – and old Books who don't take it amiss to be left abruptly, with a little business & the common cares of life fill up the sluggish hours of sensation.

We are all very well in other respects. – I am charged with making many apologies for Miss Emily's not answering Mrs Watt's kind letter just now. – A little more self-possession, and she will be able to return with gratitude her obliging sentiments and most sensible reflections.
I ever am, Dear Sir,
Yours most faithfully,
Robert Mylne[20]

The reference to Robert having been fitted out and parted for life on his journey for Gibraltar was yet another major upheaval in the family. It seems that the apprenticeship to Henry Holland three years earlier was not a success. Perhaps the work did not suit him, or vice versa, or perhaps the lure of a more adventurous life was too attractive. Britain was at war with France, Robert was eighteen, and one of his slightly older Home cousins was already an officer fighting in Spain. In March 1797 Mylne bought him an Ensign's commission – the normal way of becoming an officer – and in April took him to the military academy at Woolwich, for training at the rate of one hundred guineas a year. Three months later he received orders that meant he had to set out immediately after his mother's death and before her funeral. Four days later he embarked on a single-decked brig, the *Kingston*, off the Isle of Wight to sail in convoy to Gibraltar to join the 28th of Foot, the North Gloucestershire regiment.

That must indeed have seemed like a parting for life, just as Caroline's marriage had been. Despite the urgency of his departure, weather conditions then delayed the sailing, and on 25 September Mylne made the despairing note in his diary that Robert had only just sailed from Falmouth, after being nine and a half weeks down channel; possibly an uncomfortable experience for a boy unused to autumn weather at sea. Worse was to follow, for news then came that the *Kingston* had been separated from the rest of the convoy, and then in October *Lloyds List* reported that she had been captured by a French privateer *La Hoche* and taken into the port of L'Orient.[21]

Privateers or corsairs were as much of a risk to shipping as hostile warships. They were legal until 1856, and all the maritime nations used them to supplement their regular navies in wartime. Mylne's own great uncle James had sailed as one of Captain Clipperton's lieutenants on a famous privateering voyage against the Spanish in the Pacific in 1719. They were privately owned warships licensed to attack and capture ships belonging to, or carrying cargoes to and from, hostile nations, actions that would normally be piracy. They were often fast frigates, heavily armed and with sufficient seamen to put a prize crew on any vessel they captured, for profitable capture rather than destruction was their objective. Smaller vessels were used to prey on the coastal trade of the North Sea. They were an important weapon for the destruction of an enemy's commerce and manpower, and during the Napoleonic wars captured over 4,000 British seamen as well as seizing many ships. England had privateers too, large and small. The smallest were built in ports including Dover, Folkestone and even Hastings, where one was launched down the beach.

The capture of the *Kingston* added greatly to Mylne's miseries in the months that followed, and he must often have feared the boy was dead as he waited for each day's post. It may not be a coincidence that there is an unexplained gap in his diary about this time – the pages covering four weeks at the end of 1797 are not just blank, as often happened, but have been removed. That is something that happens nowhere else in forty-nine years of diary keeping. It was probably the unhappiest year of his life and it must have seemed that his world was falling apart.

All this time Emilia's condition was worsening. In December Mylne had taken her to his niece Lady Thompson near Southampton, on medical advice that the sea air would help her breathing. A few weeks later he wrote to Boulton and Watt about a steam engine, and then turned to family matters:

> I am at home alone – I carried Emily to Southampton a month agoe ... Harriet and Charlotte attend on her, and Willy is there also... Robert is not yet heard of since he left England ... for Gibraltar.[22]

Emilia was still at Southampton on 24 February 1798, on which day she died, at the age of twenty-four. We know almost nothing about her life. One poignant entry in Mylne's diary for 25 April 1780 captures a single fragment of her childhood, and from the sound of it may have been written as she stood watching: 'Miss Emilia Mylne brought me a shirt of her making – being 7 years old today.' That may have been a family tradition, for the only similar entry relates to Maria on 13 October 1778 and reads 'Mary brought me a Shirt made for me all done by herself – Aged 6 years, 7 months, 24 days.' Now both were dead.

Some happier news eventually came in March. Robert was alive, but a prisoner in France. Within months his return was arranged, probably by exchange for a French captive but possibly for payment. Mylne's diary records the news of his progress – first to Brest, then Paris, then Fontainebleau, and finally on 21 October landed at Dover. It had been a dreadful time. As Mylne wrote to the directors of the Gloucester and Berkeley Canal company to explain some inattention to their affairs:

> My domestick sensations have been of the most pungent kind by the loss of an eldest Daughter after two years illness, and having found a Son in a French Prison, after an anxious search for six months.[23]

Mylne celebrated Robert's return by transferring to him the house and estate at Powderhall and arranging to buy him promotion from ensign to lieutenant and for a short time what remained of the family was reunited.[24] But Robert's leave was soon used up, and he was still needed for service in the peninsula. On 17 November he set off once again for Portsmouth and Gibraltar, and this time embarked on a warship. HMS *La Minerve* was a 38-gun frigate with a crew of 300. She had been captured from the French and taken into the Royal Navy after a fierce engagement off Toulon in 1795, a common practice in a war when Britain was desperately short of ships, and had kept her old name to rub salt into the wound. She had been Nelson's flagship in the Mediterranean two years earlier, and was well equipped for the journey she now faced. In March of that year, after escorting a batch of prizes to Gibraltar she had been sent back to England with a convoy from Oporto, in order to have her bottom re-coppered and for a general refit in an English shipyard that was needed after her years of constant service at sea. Such was the press of ships awaiting work at Portsmouth that it was not finished until November, giving her captain, Cockburn, the chance 'to visit his Family & Friends whom, during this period of active Warfare, he had not seen since he parted from them a Midshipman in 1792, and indeed had scarcely been permitted to remain a week with since he entered the Profession in 1786.'[25]

So it came about that instead of the little brig of his first voyage, Robert's second was in a newly refitted and heavily armed warship, the protector of the convoy rather than one of the protected. This must have seemed a good omen and his father noted the name of the ship and its captain in his diary, yet the family's anxiety as they waited for news of his safe landfall can be imagined. *La Minerve* set sail for Gibraltar from Spithead on 20 November in company with the *Endymion* and a convoy of merchantmen.[26]

More than two months passed before news finally reached the family on 25 January 1799, presumably in a letter that has not survived. Although delayed by winter storms, *La Minerve* safely delivered the convoy to Gibraltar on 27 December before sailing on to join Nelson at Palermo. Inevitably there had been some deaths on board the various ships, as there always were. The captain's log for *La Minerve* lists her casualties, and one of the entries, made off Cape Finistere on Saturday 8 December after crossing the Bay of Biscay, simply records 'At 4 Departed this life Mr Melone, Ensign of the 28 Regm't.' No cause of death is shown, and illness seems most probable: the ship had not been in action. Robert may have shared his sisters' frailty and the weeks at sea interspersed with months as a prisoner will not have strengthened him. The spelling of his name is presumably the result of a mishearing, but his identity was not in doubt. Following the normal, all too regular, routine there will have been a simple service on deck before his body, sewn up in canvas and weighted with shot to make it sink, slipped overboard and the ship sailed on.[27] Mylne's feelings as he read the letter and prepared himself to pass that news on to his motherless younger children can scarcely be imagined.

The months that followed were full of reminders of Robert. The transfer of Powderhall had been an empty gesture, and it was advertised for sale and eventually sold. His lieutenancy had never been put through, and the money paid came back. In October came a draft for the value of his kit and belongings, which had journeyed on to reach his regiment at Minorca where they had been auctioned among his fellow officers. It was a death that must have seemed very cruel so soon after Mary and the two daughters. Five years earlier they had been a family of ten, but since then there had been four deaths and one marriage. Robert, aged sixty-six, who had never meddled much in domestic affairs, now had the care of a household that consisted of Harriet, William, Charlotte and Leonora, aged twenty-five, seventeen, fourteen and eleven. They probably made a sombre group around the breakfast table.

There was one more family death that year. It was a vital part of British naval strategy to keep the numerically superior enemy fleets bottled up in port. Not only did it remove the danger their ships might cause when out, it meant that their crews had less chance to develop the teamwork needed to manoeuvre fighting ships at sea. It was an autumn and winter spent commanding the unceasing blockade of Brest that finally broke Admiral Sir Charles Thompson's health, and he died in March 1799 within a few weeks of having come ashore (see illustration 10).

As an officer, Thompson's first ship had aptly been named the *Arrogant*, and his career was marked by a defiant independence of mind, often close to insolence, that upset senior officers but sometimes made him a very effective commander. In February 1797, as a vice-admiral, he was second-in-command to Jervis at the battle of Cape St Vincent, when his failure to see a signal earned him Jervis's fury but fortunately did not prevent the ensuing British victory, as other ships in Thompson's division, including one commanded by Nelson, saw the signal and followed it.[28] When a month later Lord Spencer at the Admiralty wrote to tell Thompson of the King's wish to make him a baronet in recognition of 'your conduct in that glorious victory', he answered that although highly gratified by the offer he would like to decline it if he could do so with propriety. He had merely done his duty following orders, did not think there was any precedent for a baronetcy in such circumstances, and 'being perfectly satisfied with His Majesty's approbation' considered that sufficient reward. The baronetcy was nevertheless bestowed.[29]

A few months later, still serving under Jervis – by then ennobled as Earl St Vincent – in the blockade of Cadiz, he had the temerity to criticise his commander's order for four mutineers to be hanged on a Sunday – not from any softness of heart, but because he thought it should have waited till the Monday. Baronet he might be, but an incensed St Vincent declined to put up with such insubordination any longer, and Thompson was sent to command the fleet blockading Brest, fell ill and died.

Mylne became guardian to Thompson's sons who were all minors. This brought his widow Jane increasingly into contact, and as she was his favourite niece they may have been some comfort to each other. The death also brought Mylne some extra work as Thompson's executor, which may have been a welcome distraction from his own sorrows. At least one of the resulting letters may even have brought a wry smile at its opportunism. Just five days after the death the Duke of Beaufort wrote, ostensibly to express his concern to Mylne at the death of his good and worthy friend Sir Charles. He felt it was too early for him to write his condolences to Lady Thompson and the children, though he felt very much for them and would write in a post or two. Then to the point – it was of course far too soon to trouble the widow on a business matter, but might he take the liberty of mentioning to Mylne his wish that 'if she intends to part with any of Sir Charles's Wines, I shall be very happy to be a Purchaser of them.'[30]

As the eighteenth century neared its end Mylne must have had many reflective moments. He was still engineer to the New River, which faced a difficult period as its antiquated network of wooden pipes became increasingly unable to cope with London's growing demands and as its competitors increased in size and number. He was also in demand as an expert witness on canals, river navigations and even steam engines, having given evidence again for Boulton and Watt in their case against Bull for patent infringement.

Yet he must have felt that his career had passed its peak, and his family life had been torn apart. If he had ever been asked who was to blame for Robert's death and Thompson's he would have settled on Napoleon, for it was his war that pitted France and her allies against Britain. It was a war that would not finish in Mylne's lifetime, and victory was by no means certain. It was a worrying time even if all had been well in the family, but it does not follow that the Mylne household was anti-French. The French revolution was barely ten years old, and France was seen as a divided nation. The girls still received their language lessons, and a Miss Godin became their governess after their mother's death and remained so for three years until she went back to France, while young William had a Monsieur Faisant as a tutor in 1798, to teach him French and Latin.

The previous year, when the depleted family was anxiously waiting for news after learning that the *Kingston* had been captured as young Robert made his first journey to Gibraltar, Mylne was busy with preparations for one of the great public occasions of the time. This was the Thanksgiving Service for naval victories at St Paul's on 19 December 1797, attended by the King and Queen. The Battle of Cape St Vincent, one of the three victories commemorated, had a particular significance for the Mylnes as Admiral Thompson had received his baronetcy for that, while Commodore Nelson's bravery had brought him to public notice for the first time as well as a knighthood and promotion to Rear Admiral. For the Mylnes and for most others Napoleon was the enemy, and the man to beat him was increasingly seen to be Nelson.

11

'Si monumentum requiris…'

How Mylne greeted the new century his diary does not disclose. Apart from the demands of family life as a widower there was much to be done on the New River, continuing work at St Paul's and he was in demand as a consultant on matters such as the future of London Bridge and the development of London's docks.

Once William left school Mylne began training him and when he was twenty-three he was taken on by the New River Company as his father's assistant. This was in 1804, and after his death sixty years later his obituarist noted that he was not originally intended for architecture or engineering.[1] It seems from this that young Robert was expected to take over from his father, as is confirmed by his apprenticeship to Henry Holland. By putting aside a separate 'fortune' for William – as he described it in his diary – Mylne was putting him in a position to be able to enter his chosen calling as a gentleman of independent means. One wonders if there had been any plans for young Robert and William to follow the footsteps of their namesakes of half a century earlier and set off across Europe. If so, Mylne clearly did not mean them to scrimp and shiver as he and his brother had done. But the war with France made travel difficult, and then young Robert's change of career and untimely death made William his father's heir and professional successor. After that he could not expect to be away for more than a few weeks at a time, and any chance of study in Europe was gone. Eventually he travelled to Paris to advise on its water supply, but that was not until the year after Waterloo, and by that time he was in his thirties with a young family.[2]

By 1800 Mylne's long connection with the Duke of Argyll was drawing to a close. His last building at Inveraray was the church, which had to cater for both the Gaelic minister and the established church. Mylne's solution was to design two identical churches back to back, making a long building with a dividing wall half way, and an entrance at each end, so that the two congregations worshipped under the same roof but in their own space and language, as they continued to do until the 1950s. He had also wanted to add a pair of semicircular colonnaded porticos, one on each side. One was to be a true porch while the other, at the head of a broad square facing the court-house, would provide a small covered market stance. This was a sensible arrangement for a town with high rainfall but the Duke had other ideas. As a result Mylne had to spend time preparing numerous designs, as exasperated notes in his diary show, none of which was ever built and this seems to have soured their forty-year relationship. Like Mylne the Duke was growing old and he may have regretted the split. When his son at Rosneath Castle was planning great changes and flirting with new architects in 1803, the aging Duke warned him against them:

> Taste without Prudence and Oeconomy, is a Mill stone about a Mans Neck, and therefore I hope you will not associate too much with Bonomi and Nasmith – You will find them expensive Pets – they will not consult your Pecuniary Interest as poor old Mylne us'd to do mine.[3]

Mylne had already transformed Inveraray Castle and that, together with his improvements to the town, and remains of a striking circular farmstead now provides the greatest surviving cluster of his work. The effect of such features as his screen wall along the waterfront was to give Inveraray a sophistication 'unrivalled in Highland towns, or indeed probably in any small town either side of the Border.'[4]

Inveraray church still stands but its appearance, and that of the town generally, was spoilt in 1941 when the spire was removed, probably unnecessarily. The area was intensively used for training special forces and the resulting heavy traffic brought fears of a collapse. An attempt to raise funds to restore it began after the war but then foundered.[5]

In England, after re-fronting Stationers' Hall by St Paul's in 1800 (see illustration 40) Mylne had almost stopped his architectural work and his diary shows how he was increasingly asked for advice on engineering matters. London's life as a trading port was being hampered by the lack of docks, and most ships had to moor in the river until they could be unloaded. This made them vulnerable to organised networks of criminals who could mingle with lawful river users and plunder their cargoes. Meanwhile the river was increasingly choked with the products of the city's unceasing growth, causing large malodorous shallows of muddy deposits, which had rendered Woolwich dockyard almost unusable. This led to his spending many days on the river with other engineers examining the sites of possible docks and seeking ways to improve the river.

He also advised, as did others, on the long overdue replacement of London Bridge. Telford wanted a single span of cast iron 600ft long, but Mylne pointed out that the brittle nature of cast iron meant that 'a small Petard exploded, when hung and attached to an important part, might, by one man in the night precipitate the whole at once into ruin', and proposed a five-arch stone bridge strong

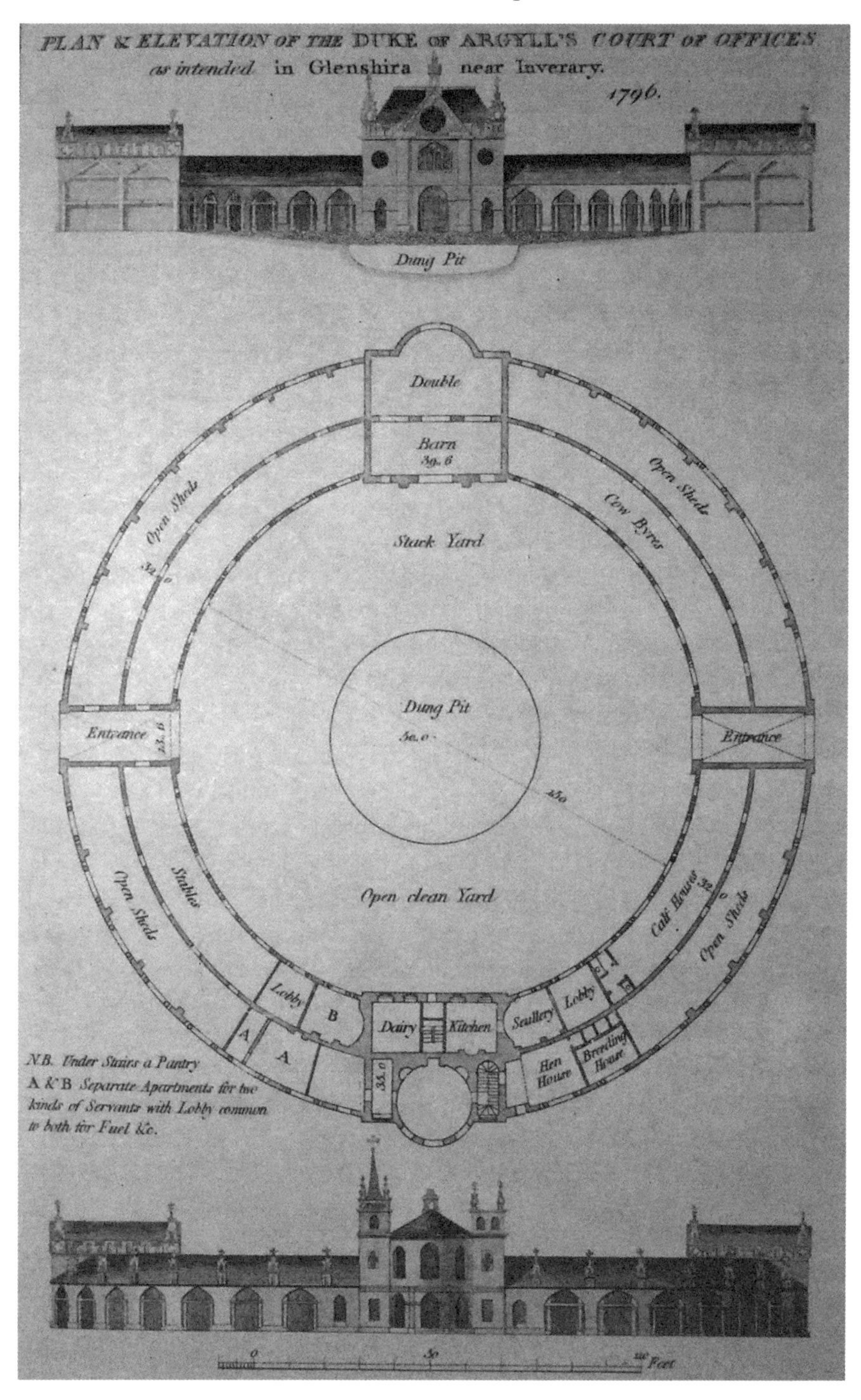

Circular farm buildings for Duke of Argyll. Among Mylne's many designs for the Duke of Argyll's castle, town and estate at Inveraray this farm building designed for a remote glen is remarkable for its mixture of functionality and striking design. Parts of it are still in use.

enough to resist 'the approach of Shipping and all the complicated Movements thereof, where raging Tides, Land Floods, floating Ice and violent Tempests are at Times to be expected, sometimes in a combined State.'

As for the surrounding area he suggested improvements including a new and level Monument Square on the City side. This would have Wren's Monument to the Great Fire at its centre, while a new circular piazza on the Southwark side could have 'the intended Naval Column' as its centre as well as improvements to the quays and a new Custom House.[6]

The gradual handover of work to William gave him time for other matters, and from 1800 his mind increasingly turned to memorials. All his life he had a great respect for stone, the material of his forebears. His original design for Blackfriars Bridge had included what would have made London's most striking collection of outdoor statuary – sixteen naval heroes gazing over the river from places of honour between the columns on the piers, protected from the ravages of weathering by the projecting recesses overhead.

At Blackfriars, in addition to the foundation stone laid on dry land for the outgoing Lord Mayor in 1760, which held one of Mylne's prize medals, the start of work on the first pier was marked with a foundation stone. That was in June 1761 when members of the bridge committee paid a ceremonial visit to the floating caisson for that purpose. That was reported in the press and then nothing more was heard of it until 1834 when the bridge was under repair and workmen went below water level in a coffer dam to examine the base of the pier. The 1761 stone was then rediscovered, having been invisible for over seventy years. It was still in good condition and proclaimed itself to be 'this the first stone of the first pier'. That seemed likely enough until a few days later the men found another stone below it, very similar in appearance. On this one, although the gist was similar the wording was different. Contradicting the upper stone's claim it proclaimed itself to be the first and continued that it 'was laid by Joseph Dixon Master Mason to this Bridge, Rob't Mylne Esq'r, Arch't.' Whatever the 'official' stone might say to the contrary, Mylne and his mason had made sure that posterity would one day learn that they were there first. They probably did it because the mason spoilt the stone with an error in the date as he cut the first line, still to be seen on the stone. They must then have decided not to waste it, but to complete and lay it for their own amusement, a mild joke for posterity to find.[7]

The first major piece of work Mylne did for the New River Company was a new office building at Blackfriars after its predecessor burnt down in 1769, and this had a foundation stone with a hidden deposit of coins.[8] Soon after he built a stone gauge at Chadwell near Ware, while he was still technically Henry Mill's assistant, which bears the stone inscription 'Mill and Mylne Engineers' and the date. When he cased the Oak Room at New River Head in new brickwork, he put inscribed stones over the windows to record the dates of original erection and reconstruction, and they are still to be seen having been transferred along with the Oak Room to the present building when its predecessor was demolished early last century. Similarly in 1778, while clerk of works at Greenwich, when he noticed that an elaborate inscription to the King and Queen Charlotte in an alcove on the river terrace had been drawn on the stone but never chiselled out he took the trouble to have it cut properly. He plainly cared about such things.[9]

This wish to leave an enduring record probably came from his upbringing. After an Edinburgh childhood surrounded by memorials to past Mylnes while being educated on the Latin classics, he had seen for himself at Rome and Pompeii how the cities the writers lived in and described had passed away. He knew of the major monuments of Egypt and Asia Minor, the product of forgotten civilizations, and he knew from the French Revolution how quickly a dynasty could be destroyed. Not only did stone buildings last indefinitely, but foundation stones could protect a hidden deposit from human curiosity for as long as a building stood, while events could be recorded for all time in stone-cut inscriptions.

The first thing he did after 1800 was no more than countless widowers and bereaved fathers have done before and since: to build a mausoleum for his wife and children. He never meant it to be his tomb, as he had other plans for that, but he purchased a plot in Amwell churchyard and built a classical mausoleum topped with an urn. There the bodies of Mary and the two girls were transferred, and inscriptions cut to their memory and that of his sea-buried son. His inscription to his wife is less formal than many, telling of her 'warm heart, accurate judgement, engaging wit and lively humour', and crediting her with 'the virtues which elevate human nature and the milder graces which adorn it.' He had already designed memorials and mausoleums for relatives and clients[10] and it cannot be said that the one at Amwell was unusual. It is mentioned here because it can be seen as the first of a series of similar actions.

The next was literally within a stone's throw, just below the churchyard on a little island he owned in the New River near one of its original springs. There was no memorial to the man who completed the New River – Sir Hugh Myddelton, so Mylne designed a handsome monument 8ft high, which is still there. It is a classical urn mounted on a square plinth with inscriptions to Sir Hugh's achievement on all sides in Latin and English (see illustration 39).

Although Mylne was still employed as engineer to the New River it was he, and not the Company who bore the cost of this substantial monument, which has lasted two hundred years and may be there for centuries to come. Mylne wrote a short description of the scene and the monument, and did so in a way that explains his feeling for the place. It describes how the New River flowed past the steep slope of Amwell Hill, gliding on either side of an islet planted with weeping willows, poplars and evergreens. It was a place 'secluded from highways and the more busy scenes of men' where the only sounds were those of the cottagers, their children and their livestock, where 'a mild and pleasing serenity steals on the mind, and soothes the senses with the effect of universal benevolence.'

Next on the list was Wren for whom he had a great admiration, as Wren's first biographer records. It was an odd fact that in 1800 there was no conspicuous public monument to London's most famous architect. His problem may have been that by surviving to ninety-one he lived too long; his reputation was already damaged by the rivalry which led to his dismissal as Master of the King's Works five years earlier, and he fell so far from favour that it was a century after his death before the first biography appeared.

Mylne has sometimes been wrongly credited with originating the epitaph *Lector, si monumentum requiris, circumspice,* that tells the reader who seeks Wren's monument to look around, but it would be truer to say that he made it well known – it was already there, on a memorial stone over his grave in the crypt, put there by Wren's son.[11]

The stone is of high quality, and Wren's biographer James Elmes thought that the injunction to 'look around' showed that it had been meant for a position upstairs in the cathedral until prevented by some cabal or intrigue.[12] The crypt at that time was very different from the present well-lit concourse with souvenir and book-shops, restaurant and cafe among the tombs and monuments. In 1800 it was used for storing all the dusty lumber of a great building, piles of wooden scaffolding and the masses of temporary seating that had to be assembled upstairs each year for the charity children's service. It had no lighting before the 1850s apart from what filtered into the side aisles through gratings, and the centre was in perpetual gloom. As one visitor to Wren's tomb pointed out, the injunction to look around in a dark gloomy vault had the opposite effect from that intended.[13]

So it was that while the main part of St Paul's was fast becoming a sculpture gallery of naval and military heroes, a practice that started in the 1790s as the war with France continued, there was no memorial to Wren in the building that dominated London except a small tablet in the dingy crypt. Mylne thought that was wrong and did something about it. In 1803 the Architects' Club put forward the proposal for a memorial to Wren in the body of the cathedral and it comes as no surprise to learn from Farington's diary that '... Mylne the Surveyor takes a lead in it ...'[14] The idea found some support, but then became bogged down in a morass of committees, while for Mylne other priorities intervened so that it was years before there was any further progress.

Eventually it ended in a compromise – there would be a conspicuous monument, but it would take the form of a marble tablet repeating in large gilt letters the Latin inscription already in the crypt. The internal layout of the cathedral was different from its present state of being open from end to end and side to side. In 1800 it had gates and a screen closing off everything east of the dome, and it was beyond that barrier in the choir that services were held. The screen carried the organ, and the place chosen for Wren's monument was on the screen above the gates, facing the main west door where every visitor would see it.

It was an important matter for Mylne, as we can judge from his diaries. He had no need to record work for which he would not be separately paid, but he regularly noted the progress of this matter. In June 1804 'Architects were with me at St Paul's on Sir Christopher Wren's statue,' which suggests the kind of monument he would have preferred. By 1807 he was able to note that he had given the Archbishop of Canterbury a drawing of the monument, and that the block of marble had arrived. In 1808 he noted that he met the Archbishop and Home Secretary at St Paul's to discuss the inscription. Not until July 1810, less than a year before his death, was he able to record that he had set the great stone of Wren's monument, and later removed its scaffolding so that it was finished and ready for public view in August. It was to be his last significant work, and Elmes later wrote that it was Mylne's fairness and justice that remedied the earlier wrong to his predecessor.[15]

That monument did not survive as long as Mylne must have hoped. In 1859 the choir screen was removed to open up the interior, and the monument moved to a less conspicuous position over the north transept doorway. There it remained until the London blitz and the night of 16 April 1941 when St Paul's suffered its worst attack. A 500-pound high explosive bomb crashed through the roof of the north transept and exploded halfway between the vaulting and the church floor, causing great structural damage. Falling masonry smashed through the floor and into the crypt, and in passing destroyed 'the marble porch bearing the famous inscription to Sir Christopher Wren'.[16] Most war damage was reinstated from public funds, but in this case the current surveyor wrote to the War Damage Commission in 1956. He explained that as the vestibule and gallery bearing the memorial tablet to Wren had been completely smashed and very little of the inscription survived, the Dean and chapter had decided not to reinstate it, but to remove a circle of marble currently in the floor under the dome and put a new inscription there in its place, inlaid in marble. Mylne might have thought Wren deserved more than an inscription underfoot.[17]

Memorials can be of paper as well as stone and Mylne recognised the historical importance of architectural drawings and engineering reports. Within months of his appointment as Surveyor of St Paul's he had located almost two hundred of Wren's original drawings made for the cathedral, which had been sold and were in the hands of a printer. He bought them for the cathedral and arranged to have them bound into volumes, so that drawings have survived where they belong that might otherwise have been lost or dispersed. A few years later he bought for St Paul's eight of Sir James Thornhill's first drawings for the paintings in the dome, which he found at a sale, as well as a perspective drawing of the cathedral by his predecessor Henry Flitcroft.[18] He then did much the same at Greenwich Hospital, with drawings by Inigo Jones and Hawksmoor, though in that case he did not have to buy them as they were lying about loose. But it was he who had them bound, which was as much as anyone could have done to preserve them for future generations. Neither at St Paul's nor at Greenwich was this part of his job, and it shows how he often had posterity in mind. He also preserved copies of all his own orders and drawings while building Blackfriars Bridge, and had them bound so that they have survived to the present day as a unique record of his greatest work.

John Smeaton's death in October 1792, which Mylne unusually noted in his diary, drew attention to the fact that his reports and papers were scattered and his life unwritten. His career had been outstanding, and Mylne and others went to a great deal of trouble to set matters right. Sir Joseph Banks bought the surviving papers, and members of the Society of Civil Engineers formed a committee to have them published. It was not done for gain, for the agreement was that they would bear the cost of publication and any loss, whereas if a profit resulted it would all go to Smeaton's executors for distribution in accordance with his will. It is no surprise to learn that Mylne oversaw the editing and in due course wrote the preface to the reports, which took years to publish. The first volume was out in 1797 but the full set of three was not complete until the year before Mylne's death.

It seems safe to conclude that Mylne believed that outstanding service merited some kind of permanent tribute, and while the Wren monument was still under discussion there had been another flurry of activity at St Paul's. It was all to do with Nelson, and deserves a chapter of its own.

12

'... the Greatest Seaman...'

From the late 1790s the public adored Nelson, partly because of his victories, and partly because of the human frailty shown by his affair with Emma Hamilton. Napoleon's threat to invade was something that only the Navy could stop, and Nelson became a symbol of the national determination to defeat him. The threat was at its highest after the short-lived Peace of Amiens broke down in 1803, and the likelihood of invasion was very real. Napoleon had 160,000 men of the Grande Armée camped along the coast from Ostend to Boulogne, making a force that Britain could not begin to match. At Boulogne a new inner harbour had been built to moor some of the fleet of invasion barges, and others were dispersed up and down the coast. In England volunteer regiments were hurriedly raised as a response to this threat, but their effectiveness as a fighting force was doubtful. The Mylnes were not immune from all this, though as usual Robert put almost nothing in his diary and we know of it only because it entailed some expense. In December 1803 there is a note of William's expenses as a volunteer, and in March 1804 notes of subscriptions to the volunteer corps, probably the Clerkenwell Loyal Volunteers, for himself and William. So Robert at the age of seventy-one and his twenty-three year-old son both seem to have taken up arms at that time of national danger.

Nelson's death at Trafalgar took place on 21 October 1805 and the news reached London on 6 November, but although the patched-up *Victory* arrived with his body off the Isle of Wight on 4 December the funeral did not take place until five weeks later, on 9 January 1806. The King had expressed his pleasure – on Government advice – that Nelson should have a military funeral at St Paul's. Nelson's own will had delicately hinted that this was also his wish, but the naval establishment always had reservations about Nelson and did not want too much public excitement. For this reason it was made clear to Nelson's brother William that the authorities wanted 'the body ... conveyed to Greenwich with as little éclat and parade as possible,' and for this reason the remains were taken to Greenwich slowly by sea, instead of attracting crowds on the much quicker journey by road from Portsmouth.[1]

Nelson's brother, a clergyman at Canterbury Cathedral, would gain immensely from Trafalgar. Whereas Horatio's peerage had been as a viscount, William who had played no part in his achievements received the next rank up, an earldom, in recognition of his brother's success, and with it a pension and the purchase price of a three thousand acre estate near Salisbury at public expense.

As to the funeral arrangements, Nelson had a coffin long before his death. This was because his friend Captain Hallowell had one made from a fragment of the mainmast of the French flagship *L'Orient*, which had blown apart at the battle of the Nile in 1798. This he presented to Nelson, who liked it well enough to keep it first on his quarterdeck, and later propped behind his chair in his cabin. He then took it back to England when he returned overland from Naples with Sir William and Emma Hamilton in 1800 and deposited it with his chosen undertakers to be kept until needed.

In his will he had asked to be buried in the Norfolk graveyard where his parents lay, unless the King had something else in mind, so there was no question of a burial at sea. After his death his remains were at first preserved on the *Victory* in a 180-gallon barrel filled with brandy. At Gibraltar, while the *Victory* was patched up for the journey home, the cask was topped up with spirits of wine for the voyage to England, and the liquid was changed twice during the journey. Eventually, on 4 December, the ship reached the Isle of Wight, and then after a delay received orders to sail to the Nore, where the Medway enters the Thames. During that voyage Nelson's remains were briefly removed from the barrel for some more attention from the surgeon. Meanwhile John Tyson, an old shipmate of Nelson and now clerk to the naval dockyard at Woolwich arranged, at his own insistence and somewhat against William Nelson's inclination, to collect the *L'Orient* coffin from the undertakers, and sailed out with it in the dockyard yacht to find the *Victory* at the Nore, so that Nelson would not have to be rolled ashore in a barrel.[2] The body was then seen for the last time, and was still in good condition. It was then simply dressed in shirt, stockings, uniform breeches and waistcoat, with one handkerchief tied at the neck and another round the forehead, before being placed in the coffin. This was then put into an outer lead coffin, which was soldered shut and encased in pine, before being transferred to Tyson's dockyard yacht for the journey to Greenwich, while the *Victory* went on to Chatham for permanent repairs to her battle-scarred hull, masts and rigging.

The yacht made its way upstream, past riverside forts firing salutes, and lines of soldiers drawn up on the banks with reversed arms as it passed the Royal Arsenal and Dockyard at Woolwich. It arrived at Greenwich about noon on Christmas Eve, and after waiting for the tide to rise enough for a landing there, what was by now a very heavy coffin was carried ashore after dark by the party of seamen from the *Victory* who had accompanied it, to be placed inside the magnificent outer coffin which had been newly made in London and then put on display in the Painted Hall. After Christmas the coffin was shown there, a mahogany case covered in black Genoa velvet and decorated all over with gilt designs formed from ten thousand gilt nail heads showing his victories and achievements, as endless lines of people queued to see it.[3]

Everyone wanted to have some part in the funeral. The King could not attend because of a longstanding convention that sovereigns did not attend the funerals of their subjects. The Prince of Wales – later to be George IV – not only planned to attend but let Nelson's brother know that he would like to be the chief mourner. William Nelson, still enjoying his dizzy rise from commoner to Earl, was overjoyed at this suggestion, as was Emma – who was not invited – but the Government would not hear of it and insisted that although the Prince might attend, if he thought proper, the chief mourner had to be an admiral like Nelson.[4] This rebuff to the Prince would probably have had Nelson's approval if he could have been consulted. When at sea with the Channel Fleet in 1801 he heard that Sir William Hamilton planned to ask the Prince of Wales to dine, and immediately wrote to Emma that if Hamilton knew what the Prince was like 'he would rather let the lowest wretch that walks the streets dine at his table than that unprincipalled Lyar. I have heard it reported that he would make you his mistress ... although I know you would send him to the Devil was he to propose such a thing, yet all the world have their eyes upon you, and your character ... is as much lost as if you were guilty.' Four days later he wrote to her again and called the Prince 'a false lying scoundrel'.[5]

It then came to the ears of the Lord Mayor that the Prince of Wales, albeit not as chief mourner, still planned to lead the procession, and he was deeply affronted. St Paul's was in the City, and within its boundaries he came before everyone except the sovereign. Some found this surprising, but it seemed he was right, so a formula was found to maintain his dignity.

Fashionable undertakers used their influence to seek a part in the proceedings. The King's second son, the Duke of Clarence, later to be William IV, found an opportunity to write to Earl Nelson about the funeral arrangements, and ended 'Having been requested by Mr Downs the Undertaker of Lower James Street Golden Square to recommend him to your Lordship for your brother's funeral I embrace this opportunity of doing so.' Sadly this royal request had to be declined. Earl Nelson wrote back to say that firstly his late brother had preferred Peddison of Brewer Street to whom he had entrusted the *L'Orient* coffin, but anyway the Home Secretary had told him the King's undertaker Mr France had the management of all public funerals, so the matter was out of his hands.[6]

When it became known that the funeral would take place at St Paul's, Mylne as Surveyor began the necessary arrangements for fitting out the cathedral, something he was well accustomed to as there was a crowded service for London's charity children every summer that called for tiers of temporary seating in the crossing under the dome. He had also fitted out the cathedral for major national events such as the thanksgiving for George III's recovery from madness in 1789 and another in December 1797, also attended by the King, to commemorate three recent naval victories and important to Mylne because of Admiral Thompson's involvement.

The spot chosen for the burial was in the crypt, precisely under the centre of the dome, possibly at Mylne's suggestion. Whereas crypt burials were normally carried round to a side staircase, it was felt that Nelson must go straight down. There was a round brass grating in the centre of the floor under the dome, but this was too small for the coffin and had to be enlarged by cutting away masses of stone.

All this was in hand even before the *Victory* reached England, and on 23 November Mylne noted in his diary: 'Set out the Works for the Sepulchre of Lord Nelson under the Cupola of St Paul's & all matters relating thereto.' The next day he noted that he wrote to the Archbishop of Canterbury regarding a sarcophagus at the British Museum, but this letter has not survived. It may have related to the design of the structure he had in mind. The following day he went to see the Archbishop and they agreed how matters should proceed. Mylne's original plan seems to have been to case the coffin in very thick Yorkshire stone with an outer layer of marble, with Nelson's name cut into it.[7]

At first it was thought the coffin would be buried in the earth below the crypt floor, as normally happened with crypt burials including Wren's, but it was then decided to place the coffin in the sarcophagus above floor level. Mylne accordingly designed substantial brick foundations, and on these his men built a box of massive granite slabs, held together with concealed bronze cramps.[8]

As the weeks went by there were more meetings to discuss how matters should proceed. Mylne proposed that as well as the grave in the crypt there should be a memorial to Nelson in the chancel. He suggested a statue on a pillar rising from the centre of the crossing floor, directly over the grave in the crypt. The choir of the church where ordinary services were held, was at that time separated from the rest of the cathedral by iron gates and an organ screen, and the place Mylne proposed was therefore outside the part where divine worship took place, and already had some naval and military monuments nearby. He was able to produce a note showing that Wren planned a monument for this spot, and pointed out that St Peter's in Rome had a structure in the same position, but the idea still found insufficient support. At the Royal Academy, where Farington took the chair in Wyatt's absence, Flaxman the sculptor piously opposed it because it would make a mortal man the principal focus in a place formed for worshipping the almighty, while Hoppner more practically pointed out that to put Nelson there would mean that no future hero could have a place of equal honour. As it turned out it was Flaxman who was eventually chosen to sculpt Nelson's memorial at St Paul's.[9]

Meanwhile public excitement was daily increasing, and it was decided that the Office of Works, under James Wyatt the Surveyor General, should take over the preparations at St Paul's, as had sometimes happened at earlier ceremonies.

Mylne will have been displeased by this. The cathedral had been in his care for forty years and none knew better than he how it should be arranged. Apart from that, Wyatt was as different a man from Mylne as could be imagined. He had every fault of Athenian Stuart and worse. The architectural historian Colvin called him an incurable absentee with the temperament of an artist and the habits of a rake. The previous year Farington noted in his diary that 'if Wyatt can get near a large fire, and have a bottle by him he cares for nothing else'. For Pugin, 'all that is vile, cunning and rascally is included in the term Wyatt'. Matthew Boulton for whom Wyatt was carrying out alterations once wrote angrily to him of how he had apparently paid him 'a very large sum of Money to bring my dwelling house into the most uncomfortable state possible, as Winds Rain and Snow drives into it ... it is constantly filld with Smoak by which my Books are spoiled my daughters health much injurd & my servants obliged to live out of Doors'. Another client William Beckford, himself no saint, once declaimed as he fumed and waited for Wyatt at the incomplete Fonthill Abbey – whose landmark tower fell down not long after Wyatt had finished it – 'Where infamous Beast, where are you? What putrid inn, what stinking tavern or pox-ridden brothel hides your hoary and gluttonous limbs?'[10]

The division of roles between Mylne and the Office of Works was such that Mylne continued to be responsible for constructing the tomb, and the location of the extra seating, while Wyatt's office installed the seating and constructed the machinery that would lower the coffin from the chancel into the crypt at the culmination of the service.

As a result Mylne may justifiably have felt that he, who had through years of devoted service earned the right to manage this affair, and who wanted to do so because of his admiration for a man he called 'the Greatest Seaman and Warrior that has ever existed' was being shouldered out of the way by more powerful interests.[11]

An illustration of the way in which public offices were jostling to be associated with Nelson at this time is the squabble over the ownership of his funeral car. The College of Arms had certain ancient privileges, which its officers had decided to exercise. On 6 January its Clarenceux and Register wrote to the Lord Chamberlain to stake a claim, 'in the Name of the King's Heralds and Pursuivants' to all velvet, broadcloth, silk, cotton, linen and all other moveable articles from the barges appointed to carry the body from Greenwich to Whitehall, requiring them to be delivered up 'as soon as may be convenient after the ceremony of conveying and attending the Corpse as aforesaid.' By 10 January, the day after the funeral, the College of Arms was even claiming the funeral car itself, so that the Lord Chamberlain's office wrote to the Garter king of arms asking on what ground or precedent they based the claim, sparking off a three-way bureaucratic tussle over a used hearse. Garter replied that they had never conceived any doubt as to their right to it, and what was more they had unanimously resolved that morning to offer it to the Admiralty for deposit at Greenwich Hospital, and further that the Admiralty had already signified their grateful acceptance. The Lord Chamberlain was not having this, and his Secretary wrote to tell the heralds 'that he does not conceive the College of Arms have any Right or Title whatsoever to the Car in Question.' No, it was his to dispose of, and the same day the Lord Chamberlain himself wrote to the Admiralty asking them to take no further steps in the matter, as he conceived the car to be at his disposal and he wished to present it to – Greenwich Hospital. The Admiralty meanwhile, ignoring both these competing claims, wrote to the Secretary of Greenwich Hospital in a letter making it quite clear that it was they, the Lords of the Admiralty who were making the presentation of the funeral car, and that it would be conveyed to Greenwich the next day, which was done.[12]

Mylne knew that though the public idolised Nelson there were many in the higher reaches of public life who did not. Some disapproved of his affair with Emma, others questioned his qualities as a commander and successes sometimes won by disobedience to orders, others may simply have been jealous.[3] Certainly no fewer than nineteen admirals, more than a third of those who were in England and had been invited, declined to attend the funeral, and they were not all ill. One who declined was Earl St Vincent, Nelson's commander at the Nile. He blamed ill-health when refusing the invitation but is known to have been out riding for three hours the day after the funeral, and had written to his sister the week before saying that Nelson's 'codicil' pleading for Emma – whom he referred to as an infernal bitch – had thrown a shade over the lustre of his service.[14]

Nelson's last action before battle had been to write the so-called codicil, though in law it was not a codicil so much as a plea for public funds for Emma and their daughter Horatia, whom he could not provide for:

> … I leave Emma, Lady Hamilton, therefore, a legacy to my King and Country, that they will give her an ample provision to maintain her rank in life… also … Horatia … These are the only favours I ask of my King and Country at the moment I am going to fight their battle.[15]

Emma had arguably done some service for the Government when at Naples in 1798, and in the circumstances of the age this was a reasonable request. Despite being widely circulated and printed in the newspapers, the Government chose not to honour it, and in due course the extravagant Emma spent her way through the moderate fortunes Hamilton and Nelson had left her and fled to France to escape her creditors, where she died in poverty at Calais in 1815.

The reality was that the public wanted a major spectacle because they idolised Nelson, as shown by the crowds who had followed him everywhere during his last visit to London. The Government also wanted one but not from any veneration of Nelson, as shown by their wish to get his remains to Greenwich 'with as little éclat and parade as possible'. Their main motive was to celebrate the victory of Trafalgar to distract attention from the news of Napoleon's spectacular victory against the Austro-Russian armies at Austerlitz at the beginning of December. The stage was thus set for one of England's great public occasions, and as usual with stage sets nothing was quite what it seemed.

At this point Matthew Boulton enters the story of Mylne's life again and we need to consider him a little.

Boulton is best known for his steam-engine partnership with James Watt, but his career went back years before that. He had inherited from his father what was known as a toy making business in Birmingham. 'Toys' were not children's playthings but small decorative manufactured objects like buckles, buttons, sword-hilts, watch-chains and snuff boxes, all of which could most efficiently be produced in a highly organised business. They were light relative to their value so could be easily distributed by the packhorses that carried goods overland before the canals were built. Such manufacture needed machines to cut, press, roll and polish, and these needed a power source. In Boulton's case his factory at Soho outside Birmingham was powered by a waterwheel whose only drawback was that production had to stop in times of drought. As a result he was on the lookout for some kind of engine, not to replace the waterwheel but to pump water from its tail race back up to the mill dam, so that the same water could be recycled to drive the wheel again in dry weather.

It was then he met Watt who had invented, but not perfected, the first true steam engine, a vast improvement on existing atmospheric engines despite similarities because of its separate condenser. Boulton, who had the necessary wealth as well as engineering skills of his own, took Watt into partnership in 1775 and their engines transformed Britain's, and then the world's, industry. When Boswell visited the factory just after they started making engines for sale Boulton told him 'I sell here, Sir, what all the world desires to have – power.'

Once the power of steam became clear Boulton began to see other uses for it, and his thoughts turned to coins and medals. Up to that time coins were still made crudely and slowly by traditional methods of stamping. Anyone who has seen a hoard of old coinage whether from ancient Rome or eighteenth-century Europe may have noticed the great variety between individual specimens. Because of the way they were made no two were quite the same. Some might be lighter or heavier, thicker or thinner. Many would be struck slightly off-centre, with part of the design missing, making it easier for clippers to pare precious metal away from the edges. The bigger the variations in a genuine coinage, the easier it became to pass off clipped coins and crude forgeries. Many of the forgeries were made in

Birmingham, which Boulton as a proud native of that town found particularly irksome. Another problem was that production was slow and therefore expensive. The result of all this was a great shortage of coinage which hampered trade, especially in the new industries which needed coins to pay workers who expected wages every week.

Mylne experienced the problem at Blackfriars Bridge. The toll collectors complained that at busy times they could not inspect every coin that was tendered, and were accordingly left with piles of forgeries – in 1775 they reported that most of the 2,158 pounds weight of copper taken in the previous three weeks was bad.[16] At first they resorted to fobbing it off on those who needed change, but when this became known they were ordered to accumulate the bad money until there was enough to be sold as scrap. This always meant receiving less than the face value, so the public was defrauded. It was a problem that plagued every business and stifled trade not just in Britain but wherever coins were used.

Boulton put his mind to the problem, and by 1790 perfected a machine to solve it (see illustration 36). Its innovation was a strong circular collar that rose from the bed of the machine to encircle a prepared metal blank just as a steam-driven press delivered the equivalent of a forty ton hammer blow, turning the blank into a coin that was precisely the right diameter and had perfectly straight sides, despite being forcibly struck and compressed between two dies. Without the collar the coin would have spread to an irregular size with rounded edges, with the collar it could not. Eventually, after years of obstruction from the Royal Mint, he won a contract to produce a new copper coinage for Britain, and also made it for Indian states, for Sierra Leone and elsewhere. He exported entire mints to Copenhagen, St Petersburg and Brazil. In effect he invented the standardised coinage now in global use. He also used the presses to make very fine medals, including one to commemorate the Battle of the Nile commissioned by Nelson's friend and prize agent Alexander Davison (see illustrations 33-34). In his lifetime the significance of his achievements was well known, and the stream of visitors to his works, including Catherine the Great, was eventually so disruptive that he had to refuse further entry.

Another of Watt's inventions relevant to these events was his letter copying press, patented in 1780 and sold by the firm. A letter was written in the ordinary way using slow-drying ink, and a semi-transparent sheet of absorbent paper was placed over it. The two sheets were then squeezed together in a press. Some of the ink soaked into the blank paper – in reversed, mirror writing and a little blurred as with any blotting paper – but it could be read the right way round from the other side of the page, and this became the copy. When Boulton first brought a sample machine to London in 1780, to demonstrate before Parliament and other interested groups like the Royal Society he was surprised to find that such a useful invention had many opponents. Bankers worried that it would become a forger's tool, and he overheard a stranger in a coffee house saying the machines should be burnt and their inventor hanged.[17]

The lives of a Birmingham manufacturer and what was then called a sea officer do not obviously run together, and one might expect Boulton and Nelson to have little in common. This was not so. Nelson took an interest in Boulton's work, and one reason may have been his love of medals. In September 1801, a few months after the battle of Copenhagen, he wrote to Lord St Vincent 'I feel myself ... as anxious to get a medal as a step in the Peerage, ... for if it be a sin to court glory, I am the most offending soul alive'. He was to be disappointed in that, for no such medal was awarded.[18] Nelson even made the effort to visit Boulton and see his factory during a trip to Wales with Emma and Sir William Hamilton in the summer of 1802, when he enjoyed a rare spell ashore, although Boulton was at that time so ill that he had to receive the visitors in his bedroom.

As for Boulton, he showed his admiration for Nelson with a costly altruistic gesture. Within weeks of Trafalgar he obtained the Admiralty's leave to produce a medal with Nelson's portrait on one side and the battle scene on the other, at his own expense, to be given to every officer and seaman who had been present at the battle – 19,000 in all – and not available for purchase despite many requests. The Naval Chronicle thought it a most honourable testimony for the 'valorous tars to carry to their wives and sweethearts', and reported that the Government had given the idea the warmest approbation.[19]

As for the link between Boulton and Mylne, they had corresponded and met both professionally and as friends for over thirty years. Thus in 1775 Boulton added at the end of a business letter to Mylne that Watt joined him 'in respectfull compliment to Mrs Mylne, on whose cheeks may the Laughs and Loves ever play for your sake', and in 1788 Mylne invited Boulton to join him on a holiday in Buxton. Boulton visiting London would dine with the Mylnes, and Mylne stayed at Soho when passing, while he and his family were also friendly with Watt and his family. Mylne had given expert evidence for Boulton and Watt in two lawsuits over the steam engine patent, calling it 'one of the most meritorious discoveries of the present time'.[20] Similarly, when Boulton and Watt were to be put up for election to the Royal Society it was Mylne who advised them on the etiquette, just as Mylne had turned to Boulton for support when he wanted a licence for what became the York Hotel. They had a longstanding relationship that combined mutual respect, trust and friendship.[21]

It was on Christmas Eve 1805 that the names of Nelson, Boulton and Mylne developed their final extraordinary connection. By this time Mylne may have feared that once the funeral was over, life would return to normal and Nelson's simple tomb in the dingy crypt would become as neglected as Wren's had been. As he was already finding with his proposed memorial to Wren, committees debated endlessly and the discussions over a memorial for Nelson might founder, just as Wren's appeared to be doing at that time. As it turned out London's main memorial, the column in Trafalgar Square, was not even started until 1839, and that with voluntary subscriptions from the public.

Mylne's precise feelings can only be a matter of speculation, but some explanation is needed for what happened next. It is revealed by an exchange of six letters, one of them now missing, between Mylne in London and Boulton in Birmingham over a period of just eleven days. There is no trace of this correspondence or its subject matter among Mylne's papers in the various archives where they are to be found. The copies quoted are among the vast collection of Boulton and Watt's papers at Birmingham. Three are original letters from Mylne, two are copies of Boulton's replies written at his dictation by a secretary like most of his mail by that time, and the sixth is a missing letter apparently in Boulton's own hand. The copies of Boulton's letters are exact, because his clerk used the patent copying press. The letters are set out here in full and with all their quirks of spelling and punctuation because, rather surprisingly, they do not seem to have been considered before.

It began on Christmas Eve 1805, the day that Nelson's body finally arrived at Greenwich and more than two weeks before the funeral. Mylne's construction of the tomb at St Paul's was almost complete, and he sat down and wrote a letter to Matthew Boulton in Birmingham:

> London, Dec'r 24. 1805
>
> My Dear Sir,
> If you have any Wish, to gratify that noble and unbounded Spirit which possesseth you, by depositing Coins or Medals, of your own formation, – along with the remains of the late Lord Session, in a Massy Stone Sarcophagus, which I have prepared for their Interment in St Pauls Cathedral; I beg leave to acquaint you, that I shall feel happy in the opportunity of performing that Service. – The Space is 8 feet 4 ins long, – 3 feet 4 ins wide – and 2 feet 6 ins Deep (Internal Dimentions) for the reception of a triple Coffin. – The Actual Flag of his Lordship under which he conquered and Died, wraped round the Coffin, will be the only thing deposited therein.
> I need not say to My old friend, how much I am,
> Yours ever & most faithfully,
> Robert Mylne [22]

On the outside of that letter, in a different hand, is written 'Buryed on 7th Tuesday, place prepared 6 Monday, sent on 5 Sunday or Saturday 4th'. This presumably sets out what Boulton believed the timetable would be, and not until 28 December was the date of the funeral finally set for Thursday 9 January.[23]

The reference to the non-existent 'Lord Session' – and the word certainly looks more like Session than Nelson – is unexplained. All the rest of the letter and the correspondence that follows plainly refers to Nelson's funeral and his triple coffin, which was indeed to be wrapped in the flag under which he had conquered and died. Who else ever had a triple coffin? Presumably it was either some private pun whose meaning is now lost or, more likely, an absent-minded slip of the pen.

London, Decr. 24. 1805.

My Dear Sir,

If you have any Wish, to gratify that noble and unbounded Spirit which possesseth you, by depositing Coins or Medals, of your own formation, along with the remains of the late Lord Nelson, in a Massy Stone Sarcophagus, which I have prepared for their Interment in St. Pauls Cathedral; — I beg leave to acquaint you, that I shall feel happy in the opportunity of performing that Service. — The Space is 8 feet 4 Ins. long, — 3 feet 4 Ins. Wide, — and 2 feet 6 Ins. Deep — (Internal Dimensions,) for the reception of a triple Coffin. — The actual Flag of his Lord Ship under which, he conquered and Died, wraped round the Coffin, will be the only thing deposited therein. I need not say to my old friend, how much I am — Yours ever & most faithfully

Matw. Bolton, Esquire. — Robert Mylne

Above and below: Letter from Robert Mylne to Matthew Boulton, 24 December 1805, and an extract from Boulton's letter to Mylne of 31 December 1805. (Courtesy of The Birmingham Assay Office)

My dear Sir

If you expect me to be a punctual correspondt. you must excuse me if I employ the hand of my Amanuensis, for I am still unwell & writing always makes me worse, besides I am so busy exerting myself in selecting the coins & medals to be deposited by your hand in the Sarcophagus of our departed Friend the illustrious Lord Nelson. but I am apprehensive all my efforts will be vain unless you attend to all the previous minutiae

A Lord *of* Session is a Scottish judge as Mylne would know, but judges do not generally conquer and die under any flag, nor was one about to be buried at St Paul's. In any case the later letters make it perfectly clear who was meant.

There was an efficient post between the two cities, and Boulton's reply was evidently written the next day. Unfortunately it is missing from the collection now at Birmingham, presumably because Boulton wrote it himself and it did not go through the copying press. All we know of its existence comes from references to it in Mylne's second letter, written in reply to it on 27 December.

London, Dec'r 27. 1805.
My Dear Sir,
It was with much pleasure I recognized your hand writing, in yours of Christmas Day. Mr Lawson shewed me lately a letter of yours, not in <u>your</u> hand writing, which affected me greately. – Thank God you are likely to weather this Gale of a very hard Winter.

My little and late <u>jeux de esprit</u> was to call forth some more ebbulitions of that Noble spirit; which I have bowed to, and admired for Many years.

It was not for the purpose of any new formation or fabrication in the Medalick Art: But for a compleat Series of all you ever have done in that way, even to farthings. – And I mean to place them in a <u>thin space</u> of an Inch or so, which will be under the Coffin. – It was to bury your Glories for the instruction & admiration of future times, what was done in this Country in <u>these</u> times; Along with the Glories of the Greatest Seaman and Warior, that Has ever existed, since recording such transactions has ever existed.

It was coupling <u>you two</u> together, that induces me to think, you may embrace this opurtunity, of handing down to latest times the Merits of my <u>friend</u>, with the man & <u>hero</u> whom I <u>adore</u>. All your Coins and Medals (if you can hastily muster up one of each) will not go far to fill a Space 8 feet long, 3feet 4in broad and an Inch thick; and to be buried in Dry White Sand, so as not to be Seen.

If this idea should fall in with the Correctness of your judgment put it in my power before the 7th or 8 of next Month.

Best respects to your Son, and wishes to you both, for happiness in this world and the life everlasting – is the prayer of, my Dearest Sir, Yours
Robert Mylne

Mathew Bolton, Esquire
I could not expect, what your Spirit is now at work on could make any part of this proposition.[24]

The postscript probably refers to the Trafalgar medal that Boulton was already known to be working on but which would not be ready for some months, and which Boulton may have referred to in the missing letter.

Boulton's reply to Mylne, undated but written on New Year's Eve as we can tell from his subsequent letter, is preserved as a faint copy on thin paper from the copying press.

My dear Sir,
If you expect me to be a punctual Correspond't you must excuse me if I employ the hand of my Amanuensis, for I am still unwell & writing always makes me worse, besides I am today exerting myself in selecting the coins & medals to be deposited by your hand in the Sarcophagus of our departed Friend the illustrious Lord Nelson, but I am apprehensive all my efforts will be in vain unless you attend to all the previous minutiae as well as to the act itself of depositing the Medals, which I at first intended should be ... in a Glass Bottle; but I perceive you have only allotted for them a space of one inch thick. I therefore now propose to have them enclos'd between two pieces of plate glass with a frame made of dry thick slate, marble, or other durable material for the purpose of excluding the air of the stone, for you may be assured the Medals will not endure long if that has the least access to them. When this frame is made I wou'd have one piece of the plate glass dropped into it & the medals placed thereon as described on the Paper herewith.

The frame should then, I conceive, be filled with pulverized glass & the top plate of glass be laid upon it by which means the Medals will be wholly excluded from the air.

I propose to send them by Coach on Saturday next so that you may have time to consider the subject and make the necessary arrangements previous to the 7th.
I am Dear Sir, with sincere regard,
your faithful & ob't Serv't
Matt'w Boulton.

Below the text is a simple sketch showing how coins could be placed flat between two sheets of glass resting in an L-section stone frame. [25]

The next letter, another faint copy from Boulton's press, was dated New Year's Day, and shows him enthusiastic as ever about the scheme.

Robt Mylne Esqr
My Dear Sir
Judging it best to take time by the Forelock I delay not, as was proposed in my Letter of yesterday, the forwarding of the Medals & Coins till Saturday but instead thereof shall send them by Mail Coach this Day to the care of our Agent Mr John Woodward No 13 London Street who will be requested to send a person with the Parcel, directed to you, to New River Head the Moment it gets to hand. I shall inclose in the same parcel a sample of pounded Glass for the

purpose of securing the Medals & Coins as suggested to you in mine of yesterday, but as I cannot ascertain the quantity that will be required I must get you to have the deficiency supplied in Town.

Of the Medals and Coins sent I would have you deposit what you think proper & keep what you think proper.

Pray be careful that all moisture be kept out of the Sanctum Sanctorum. The Copper Boxes are an excellent preservative of Medals in this World; but perhaps the pounded glass may do without the boxes for the other World. However I leave that to your discretion.

The principle of the preservation of Metals is perfect exclusion from air or moisture. You will want two pair of Glass Plates, viz one for the gilt pieces & another for the bronzed. When you have placed the Medals & Glass Plates in the Sarcophagus you must take care that they are not displaced by putting on the covering stones, nor suffer any wet mortar or other moisture to enter.

Also take care that the dead are not robb'd by the living.

I remain with much respect, Dear Sir,
your faithful & obdt Servt
Mattw Boulton[26]

The 'copper boxes' he refers to may be the little round cases, in two halves made to fit snugly round one coin, which can be seen with some examples of Boulton's coins at the British Museum. Finally there is the reply Mylne sent him on 3 January.

London
Jan'y 3d. 1806.

My Dear Sir,
I received yours of the 1st Instant, and the parcel of Coins, and Medals, agreeable to the particulars, yesterday. – I have ordered 2 Plates of Glass, and a frame to place them according to the instructions, as Accurate as possible and the time will allow of. – I intend to place them under the head of the Coffin, and in the mean time to say as little as possible; for the avidity of the many headed Monster, (the Publick) are ready to tare every place to peices to have a peep.
God Bless you in the prayers
of yours ever
Robert Mylne
I shall write fully when all is over.[27]

There the correspondence ends, yet the unfinished story it tells is astonishing. Two respected public men had agreed in writing to make a secret deposit of treasure in the grave of a hero they revered, immediately before his interment. Apart from the fact that it was to be done in secret, it resembled the impulse people have to throw some last offering into a loved one's grave before the earth goes in to fill it. It has a certain audacity about it, because the entire population from the King to his most humble subject who eagerly followed every detail of the funeral would remain in ignorance of this extra tantalising ingredient.

Because there is no list of the coins and medals we cannot begin to estimate the quantity involved, let alone its value, but Boulton was a wealthy man. He also, like Mylne, had a vast admiration for Nelson, as shown by his decision to mint almost twenty thousand copies of his Trafalgar medal and present them to every sailor who had been present with the British fleet at the battle. There is no reason to think that the selection he spent a day making from his private collection would have been anything other than generous. He had many to choose from, such as those struck for Admiral Howe and the Glorious First of June, the Battle of the Nile in 1798, the re-establishment of the King of Naples 1799 and even the Peace of Amiens in 1801. He habitually struck coins and medals in a variety of materials from gold and silver down to pewter, and it is a pity that he did not keep a copy of the 'particulars' that Mylne mentions.

What is most remarkable is that not only did the plan apparently go ahead, but also that it seems to have remained undescribed until now.

What does it tell us about Mylne? One can see that he may have thought the establishment was more interested in glorifying itself than Nelson, and might do little for his memory once the excitement had died down. His own natural wish to play a part in the proceedings was to some extent diminished, when Wyatt of all people was asked to do things Mylne could easily have done, and which were clearly within his remit. Moreover, Mylne's suggestion as to the form of a monument had not found favour. As we know from Greenwich and from Wells, Mylne generally took steps to remedy anything that he felt reflected on him.

For Mylne the letters also reveal a more important reason – his regard for posterity. As he explained in the second letter, his purpose was:

> ... to bury your Glories for the instruction & admiration of future times, what was done in this Country in these times; Along with the Glories of the Greatest Seaman and Warrior, that Has ever existed.

He clearly had a belief, shared by others at that time, that the greatest civilizations eventually pass away and that one day people might stand among the ruins of London wondering what sort of people had lived there. Thus Walpole had written how future travellers from Lima might come to dig among the ruins of St Paul's.[27a] He wanted to leave them at least a hint, something to find and wonder at, as he had done in Italy, and where better to place a future archaeological deposit, a time capsule intended to be found, than under the precise centre of the city's biggest building, whose location would be made prominent if only by the tumbled remains of the massive masonry piers that upheld Wren's dome?

The coins and medals deposited were manufactured to such a degree of technical precision that they surpassed in that respect any similar items that might be found from earlier ages. Moreover the little scenes they depicted provided pictorial information about the state of Britain at the turn of the eighteenth century, giving them meaning even to people who might not be able to read their lettering.

Even after he had agreed his plan with Boulton, Mylne made a further effort to contribute to the ceremony, and it was something Nelson would have thanked him for. On 30 December he wrote to Nelson's executor Haslewood with a suggestion. He had been told of a piece of Italian poetry celebrating the Battle of the Nile and set to music. Nelson had heard it at a feast held by the King of Naples at Palermo, and had been heard to say that he would like it played at his funeral. This he had mentioned when last in London. Mylne suggested that he saw no reason to suppress it, and it could be done either in the choir of the church or afterwards in the chapel in the crypt. Haslewood must have passed this on to Earl Nelson, who put the suggestion forward to a meeting of the Dean and chapter at St Paul's on 4 January. Pridden and Page, two of the cathedral clergy, were accordingly delegated to examine the music, but predictably deemed it to be 'of too light a nature to be performed upon so solemn an occasion', notwithstanding the wishes of the deceased, and so it was never played.[28]

If Mylne had already taken the view that Nelson's own wishes were being subverted to other ends this decision can only have confirmed it and increased his resolve to press on with his own private tribute.

By the New Year demand for tickets was so great that more and more seats were constructed until Mylne said they could be built no higher. There were probably seven thousand people seated.[29] A commentator noted that while the preparations were underway 'door money was demanded as at a puppet show', so that the public were obliged to pay for a sight of the preparations which would themselves be paid for at the public expense. The four vergers, he reported, were said to have made more than a thousand pounds from this source.[30]

Minor squabbles continued. Wyatt's men had constructed the machinery for lowering the coffin into the crypt so that it protruded upwards with a gantry like a great obelisk. The College of Heralds and the Choir complained it would block views inside the cathedral, and they were made to remove it and devise a smaller one. Similarly there was a dispute between Mylne and Wyatt about the pulpit. Wyatt wanted it out of the way, whereas Mylne – who had only recently designed it and had it made – thought it better where it was. Wyatt won that one, when the Dean and Chapter realised that its removal would improve the view of the ceremony from the seats allocated for them and their friends.[31]

The funeral took place on 9 January. The previous day the coffin was taken by water from Greenwich to Westminster, and left overnight at the Admiralty (see illustration 37). On the 9th the procession was so long that the front reached St Paul's before the end had left the Admiralty, and the coffin was borne on a vast funeral car shaped to resemble the *Victory*. At St Paul's, where the service would continue after dark, the interior was lit with a lantern in the form of a cylinder 10ft high, hung round with one hundred and sixty oil lamps. By this light the vast congregation, including various Mylnes (see illustration 35), watched a three hour service, ending with the apparent descent of Nelson's coffin into the grave, after which the party of seamen from *Victory* who were supposed to drop the large flag after it could not resist tearing it to pieces to take souvenirs. This final moving scene made a vivid tableau dwelt on in all the contemporary accounts.[32]

In reality there had been a change of plan and the hero's remains were not to be at rest just yet, for reasons that were largely mercenary.

On 21 January, almost a fortnight after the coffin's mechanical descent into apparent eternal rest in the crypt, 'a lover of decorum' wrote anonymously to the Times to protest at what he called the disgraceful exhibition of Lord Nelson's coffin as a public spectacle at St Paul's. It seemed that those who collected money from sightseers had managed to obtain the Dean's approval to delay the final part of the descent of the coffin into its sarcophagus, and that it was still visible – on payment. The writer believed a large sum of money must have been collected, and asked where money so disgracefully obtained, by what he called an indecency scarcely credible in a civilized country, was destined.[33] This continuing exhibition explains the only two relevant entries in Mylne's diary that month. On the day of the funeral he wrote 'Lord Nelson's funeral but no Interment', and the next day 'Coffin 7 feet from the floor'. As the waiting sarcophagus was less than 5ft high, it seems that the coffin was still resting on its bier having been lowered from the floor above by the 'invisible machinery' at the end of the service – assumed by those who described the ceremony to be its final descent into the silence of the grave.

At this temporary resting place it was clearly visible to sightseers above, through the opening cut for its descent, and they had no need to enter the crypt for that purpose.

It was not until some weeks after the funeral that Nelson's coffin was finally inside its tomb with the lid on. Of the actual date there now seems to be no record. The delay is unlikely to have made any difference to Mylne's plan, which must always have been that his private deposit would be hidden in the bottom of the sarcophagus at some stage before the coffin's arrival. Once the deposit was there the top sheet of glass could easily be concealed with a layer of sand, stone or mortar: there was plenty of space and Mylne of all people had the necessary stonemason's knowledge.

The final embellishment of Nelson's rather austere tomb, which Mylne had intended to case in marble, was then left in Wyatt's hands. He eventually produced a selection of designs in 1807, but these found no favour and do not seem to have survived. Canova, in Italy, also proposed a design, but the Treasury consulted the Committee for the Inspection of Models, who recommended the use of an existing sixteenth-century sarcophagus then in the burial house at Windsor Castle. It was a fine piece of renaissance workmanship by Benedetto Rovezzano that Cardinal Wolsey had commissioned to decorate his own tomb in due course, and which had become Crown property on his downfall under Henry VIII, since when nobody had known quite what to do with it. The Treasury approved this in February 1808 and instructed Wyatt accordingly. The result was that first a plinth of black and white marble with classical mouldings and the lettering 'HORATIO VISC. NELSON' was put on the top of Mylne's plain granite. The Wolsey sarcophagus was then added to the pile, and on top of it was placed a marble cushion surmounted by a viscount's coronet (see illustration 38). This multi-tiered layer cake structure has sometimes caused confusion about where precisely the coffin is, but original documents and reliable contemporary texts confirm that it is where it has always been, inside Mylne's rough granite base.[34]

Precisely when Wolsey's sarcophagus, and the black and white marble plinth on which it rests, were added as decorative flourishes to the granite base is difficult to discover. It was clearly more than two years after the funeral, for it was not until then that the plan was approved. It would not have called for any particular ceremony, and the event does not seem to have been noticed by *The Times* or the *Naval Chronicle*, nor is it mentioned by any of the standard published works on Nelson. Probably by the time Wyatt got round to it Nelson's memory was so overshadowed by later events that it was not thought of sufficient importance.

Wyatt's famous dilatoriness only complicates the problem. It was not until April 1813, just a few months before death in an overturned carriage brought him a merciful release from his looming problems of encircling creditors and a heavily pregnant housemaid, that he wrote to the Treasury asking for £1054 2*s* 8¾*d* for his work to Nelson's tomb. As he had not troubled to list the tradesmen to whom this was due or the amounts claimed by each there was further delay, and a year later the Office of Works was still asking staff at Windsor for duplicate copies of bills for packing cases and the cost of sending the sarcophagus to St Paul's, which they claimed to have already given to the late surveyor-general. When a slightly smaller sum was eventually authorised it referred to work on Lord Nelson's tomb up to 1810, so that is probably the year when Wolsey's sarcophagus and the cushion and coronet were added to the grave. It was certainly there by July 1810, when a visitor wrote a detailed description and thought it 'most elegant and proper'.[35]

There was to be a much more conspicuous monument to Nelson at St Paul's, a piece of monumental statuary by John Flaxman upstairs in the transept, and this diverted attention from the tomb. It is a typical piece of its time, topped by the figure of Nelson in uniform with one empty sleeve, backed by an anchor stock and a coil of cable and guarded by a crouching lion, while a helmeted Britannia gestures up at him as she puts a protective arm around two boys in sailor suits. It was simply one of a crowd of monuments being planned at that time as the bare interior of St Paul's was turned into a pantheon for naval and military heroes. In June 1806 the Royal Academy sent a report to the Treasury dealing with monuments for Nelson, Lord Cornwallis, and Captains Cook and Duff as well as the recently dead Prime Minister Pitt. Commissions were lucrative. The Academy had chosen a design by Westmacott, but the Nelson family asked that Flaxman should be the sculptor, and he was awarded the £6,000 commission. For his part Mylne asked for and was paid £25, with a further £5 for coach hire and messengers, for his work designing and supervising the construction of the sarcophagus. The cost of labour and materials for his craftsmen was a further £370.[36]

Since Mylne's time there have been a number of changes to that part of the crypt. When Wellington's burial was being planned in 1852 there were complaints that it was 'forlorn, dismal and dirty', and that it was quite inappropriate for Wolsey's sarcophagus to be over Nelson's remains. One suggestion, from *The Builder* magazine was to remove the empty sarcophagus and surmount Nelson with a monumental tomb for Wellington, 'thus putting together the two great commanders in a place of equal honour.'[37] As it happened Wellington's coffin did reside on Nelson's tomb for a time, when it was lowered there at his funeral and remained for more than a year until a sarcophagus was ready further along the crypt. Perhaps as a result of the complaints the floor level was lowered in 1855 and the old flagstones replaced by nine-inch black and red encaustic tiles by Minton, set diagonally. This was during the surveyorship of F.C. Penrose, and the specification provided that Nelson's tomb was to be left untouched except to the extent that it was to be given a stone kerb, presumably to hide foundations revealed by the change in level. The tiles may have seemed rather commonplace, and when work began to upgrade the crypt in 1874 Penrose designed the present mosaic floor, decorated with anchors, tridents, dolphins and a crocodile which took several years to complete. He thereby provided some useful toil for female prisoners at Woking who made the thousands of intricate little tesserae, the Superintendent of Prison Discipline having offered their output on 'very advantageous terms'. More recently, in 1956, the Admiralty paid the costs of a general clean up and regilding of the tomb and sarcophagus, which had escaped damage during the blitz but bore the grime of a hundred and fifty years.[38]

Despite these changes to its surroundings, it seems there has been no interference with the tomb itself since its completion two centuries ago. There is no reason to doubt that Mylne carried out his plan, for we have seen that it was not his habit to give way once he had decided on a particular course, a characteristic he shared with Nelson. By the time of the funeral the cathedral's structure had been in his care for almost forty years, and nobody could challenge his authority to come and go as he pleased, at any hour, anywhere in the vast building including the darkness of the crypt, where he was in any case well known to be directing the construction of the tomb. Everything we know about his character and his assertive manner suggests that he would have had no difficulty in excluding anyone he did not choose to have present for the few vital minutes of deposition and concealment. If he wanted the help of a trusted workman or assistant he had the contacts of a lifetime to call on. Once the deposit was in place nobody would have had any reason to suspect that there might be something hidden in the deep recess of the new tomb, while the perpetual gloom of the crypt would have helped to conceal his actions. The three components of coins, glass and frame could be carried to the crypt separately, to be put in place at a convenient moment. He had told Boulton the space in the sarcophagus was 8ft 4in long, 3ft 4in wide and 2ft 6in deep. The undertaker's bill for the outer coffin describes it as a strong mahogany coffin case 6ft 8in long, 2ft 2in wide at the broadest and 1ft 7in deep.[39] This left almost a foot of spare depth, of which Mylne's secret deposit need not have taken up even a quarter, and so there was no shortage of space.

Mylne's drawings for the sarcophagus seem to have been lost, but more than forty years earlier he had given detailed orders how a cavity for coins in the foundation stone at Blackfriars Bridge was to be constructed and protected with clamps and dovetails, and had left a marginal note explaining that this was done '...in order to preserve the edges of the fifth stone from being cut, for otherways some people might attempt to take out the coin.' He knew how such things should be done.[40]

On the other hand it is impossible to be certain that the plan was carried out as long as there is no trace of, for example, the letter Mylne said he would send Boulton after the funeral, or some other note. That absence is not altogether surprising, as the matter was obviously confidential. Boulton's letter to Mylne had warned him to guard against the risk that the living might rob the dead, and Mylne echoed the risk when he wrote of 'that many headed monster the public' tearing things apart. This was prophetic, considering the way Nelson's flag was torn. So any such letter from Mylne may have gone straight into Boulton's fire once it had been read – and that may be the reason why the originals of Boulton's letters are not to be found among Mylne's papers, while Mylne very rarely kept copies of his own letters. Boulton by this time was very ill, and it is perhaps because of the highly organised nature of his massive business that his secretary followed the usual practice and made copies of the letters he wrote to Mylne at Boulton's dictation, just as he filed away Mylne's originals, where they have remained for two centuries with their significance apparently unnoticed among so many thousands of other papers. Had he not done so the whole curious story would have been lost.

This too is another puzzle. Why has such a clearly documented plan apparently passed unreported for so long? There can be no doubt about the authenticity of the letters, which are part of the vast collection removed from the Birmingham works to Tew House in Oxfordshire by Boulton's son in 1850. There they remained until they were passed to Birmingham Assay Office in 1921, after which they were catalogued and eventually passed to the Birmingham City archive. The head of the Assay office at that time apparently believed that some of Boulton's medals had been buried with Nelson, but whether he did so from these letters or from some other source is unknown, and the fact seems to be nowhere reported in all the reams of publications about Nelson's death and burial over the past two centuries.[41] One possibility is that the substitution of 'Session' for Nelson in the first letter, whether deliberate or accidental, has misled casual readers. It occurs in Mylne's opening letter, much easier to read than the faint and blurred copies from Boulton's copying press, which are tiring to read. Another possibility is that those who have seen the letters have been focussing on engineering history, and may have incorrectly assumed that the matter was already well known to historians of Nelson.

All one can say is that there is probably a small treasure trove of Boulton's best coins and medals beneath Nelson's triple coffin in the tomb. But unless some further document turns up to take the matter further, it seems unlikely that we can ever be sure: the tomb is sealed, deserves to remain so, and is in a location where it almost certainly will. Perhaps Mylne's deposit will stay there until the remote future he anticipated, when people come to pick over the overgrown remains of a ruined city where those things that were not hidden below ground in indestructible granite boxes will have long gone.

This raises a further question: is anything else deposited with Boulton's coins and medals? Was Mylne, the descendant of master masons who cared about posterity, tempted to add anything of his own to the interior of the tomb? It might have been no more than the mason's mark his forebears left on the structures they built, six straight chiselled lines making one M inverted over another. We will probably never know. Nor can we know if Mylne ever reflected that the word crypt comes from *kryptos*, the Greek for hidden or secret.

13

'...mine has been a very active life ...'

After the bustle of Nelson's funeral Mylne's life resumed its usual pattern and he never had a period of true retirement. His forty-ninth, and last, diary is for 1810 and shows him still active at the age of seventy-seven. That year he finished editing Smeaton's reports and sent them to the publisher, a three-volume work that runs to 1,300 pages. In May he took the New River directors on their annual inspection of the river and later attended their midsummer dinner. In June he fitted out St Paul's for a special service for the Lord Mayor and judges. He was still busy there with Wren's monument and completed it in August when he also made a survey of the cathedral. He attended meetings to arrange the maintenance of Blackfriars Bridge and other works. He gave evidence to Parliament about Thames improvements and Aberdeen Harbour. He made a report and plans for almshouses at Ledbury and advised the Duke of Northumberland on work at Syon. He attended dinners and meetings of the Civil Engineers, the College of Surgeons, the Royal Society and the Stationers' Company. He appeared at lawsuits as a witness. He met George Dance to consider plans for a new London market to replace Smithfield. In November he formally retired from the New River, with a pension as thanks for his 'long and faithful service', and his son William Chadwell Mylne stepped into a post he would hold for fifty years.[1]

For 1811 there is no diary and just after midday on Sunday 5 May he died in his bed at New River Head, aged seventy-eight. The only account of those months comes from Robert Smirke, who attended the funeral and later told Joseph Farington what he knew. Mylne had stayed at home since the previous Christmas, having no illness but being disinclined to go out. Then, shortly before his death he returned to bed after getting up one morning, and stayed there weak but without pain until the day before his death, when he was in pain from phlegm that he could not bring up, yet stayed fully conscious until he died.[2]

His will, made eight years earlier, asked that he should be buried in the crypt of St Paul's and this wish was granted. His gravestone there tells of three of his main achievements: Blackfriars Bridge, the maintenance of the New River for over forty years, and his surveyorship of St Paul's before ending 'His remains now repose under the protection of this Edifice which was so long the Object of his Care'.

He had been planning to write his own autobiography '... mine has been a very active life in conceiving and executing: it remains for its wane to register and publish some accounts of its transactions.'[3] Farington describes seeing an account Mylne had written of his life and sent to George Dance a year before his death, but does not summarise it. Dance was about to publish a book of sketched portraits, including one of Mylne (see illustration 5), and each has a page of biography. If this is all Mylne wrote it adds little to what was already known, though it does report that his youthful ambition was to honour his forebears by following their profession, and that in Rome he made a critical study of both ancient and modern buildings as well as its water supply aqueducts.[4]

After his death an obituary in the *Gentleman's Magazine* claimed that he had written 'curious memorials' of his travels in Sicily, illustrated with his own sketches, which were among his papers at the time of his death and were expected to be published in due course, but that seems to be their only mention and they are not among his deposited papers.[5]

From all we know of Mylne he must have wanted his achievements recorded, but he may have assumed it would be done after his death. This is implicit in his will of 1803, which left monetary legacies to his daughters, while the bulk of his estate went to William. The will specified some categories of goods that William was not to 'sell or give away upon any account or under any pretence whatsoever ... except in the way of publication ... and to such of the said articles as may be deemed proper for publication'. The list included his books, papers, drawings, manuscripts, models, instruments and other professional and scientific articles, as well as his 'natural curiosities, coins and medals'. Little is known of his collections but his diary occasionally records such purchases as Sicilian antiquities or natural curiosities.[6] Whether he expected that William would undertake the task, or that his professional colleagues would, as he had done for Smeaton, we cannot know. In the event he outlived most of them and it was never done.

At the end, what do we know of Mylne? Enough portraits survive to show us how he looked. From the earliest drawn in Rome to one in old age they show a man with regular features and a rather heavy jaw, and some show prominent eyes (see illustrations 3-5). He wore his hair long in a pigtail throughout his life, with or without a wig, and one gets the impression of a tall, substantial man. One portrait that he did not like was Vangeliste's 1783 engraving from Richard Brompton's drawing made in Rome a quarter of a century earlier, and characteristically he did not hide his opinion from his friend Whitefoord who had commissioned it for him.

> Among the Mylne family, who have been accustomed to the Original it is really awkward in this respect ... The Head seems not to have Brains enough; The Scull backwards is too scanty. The fillet, across the hair, does not lye straight

> ... the engraving is coarse on the face ... The nose is too long ... There is something wanting of the eye, in its expression, ... the Ball of the Eye is not round. The under Eye Lash is very stiff. The chin projects not far enough; and too much is allotted to the Under Lip.[7]

If this sounds harsh it must be said that a glance at the original – not illustrated – supports his criticism. The last portrait (see illustration 5) is to be found in a book of portraits by George Dance the younger published in 1814. Mylne appears first in the book, and this must have been Dance's choice, perhaps reflecting the long friendship that is said to have linked the two men since they met in Rome sixty years earlier. Like many of Dance's sitters he is shown in profile sitting in a straight-backed wooden chair, and drawn from a lower viewpoint that emphasises the heaviness of his jaw. Dance wrote in his preface to an earlier selection that his purpose was to record faithful resemblances of distinguished men to show 'how surprisingly nature has diversified the human countenance', so the portrait is likely to be accurate. It shows Mylne with features a little too heavy to be handsome, unsmiling – like all Dance's portraits here – but not humourless. His head is held high and a little stiffly, whether from his rheumatism or because his friend had told him to stop talking and keep still we will never know.[8]

As to his lifestyle, Farington revealed a spartan streak when he wrote that Mylne's office at New River Head was situated over water, very damp and without a fire even in winter, and that he was so careless about his clothes that he would put on a shirt 'saturated with dampness'. By contrast he wrote from his own knowledge of Mylne, who had been a friend and dinner guest, that he was a man much disposed to conversation who drank wine freely at and after his meals, and was extremely exact in all his affairs, which he noted and lotted with great care.[9]

What of his character and abilities? Over forty years earlier the artist James Bruce of Kinnaird, away exploring the antiquities of the Syrian desert, wrote that he had never doubted Mylne's abilities 'as I thought him sure of his principles; other architects are paving streets, while he is employed in bridges and palaces. There is a particular Providence presides over the arts, which, with a little patience, puts everybody in their proper places.'[10]

Lord Mansfield, one of the greatest of English judges, characterised Mylne in a 1781 judgment as 'a man very considerable in his profession, a good architect, a good mathematician and a good engineer. There is no doubt he deserves the character that has been given to him.'[11]

Mylne's problems at Greenwich suggest that he was a difficult colleague, but this is not borne out by his work for the New River Company. That required him to deal with everyone connected with the company, from the walksmen who patrolled the New River maintaining forty miles of open channel to the wealthy shareholders who took a close interest in the state of the works, partly because of the effect it could have on a dividend that fluctuated greatly. Mylne, who could have been dismissed at any time, was kept on until a few months before his death when he was awarded a generous pension and allowed to put forward his son as his successor. Earlier, in 1806 after the board members had ridden the length of the New River, an annual event since 1613, they recorded how, having found the banks, machinery and everything pertaining to the works in the most perfect repair, thanks to Mylne's diligence and attention, they wished him to accept a piece of inscribed plate as proof of the genuine good opinion they had of his integrity, talents and zeal. That too suggests respect and a genuine regard.[12] The problem at Greenwich was almost certainly the result of his insistence on high standards among colleagues who were more casual and his wish to preserve the limited building budget for repairs to the fire-damaged chapel, rather than the works they wanted in their private apartments.

He disliked any kind of waste. At Blackfriars he invented reusable centring to reduce the amount of timber needed, while the bottoms and even the sides of the wooden caisson ended up as part of the pier foundations. At Greenwich he saved a large amount of architectural salvage for the future use of the Hospital and was furious to hear later that Stuart had disposed of it. At New River Head it was he who introduced a waterwheel of his own design that cost nothing as it pumped water uphill, reducing the work of the coal-burning engine.

There is little doubt that he had a fiery temper. James Elmes who visited him many times while researching the life of Wren said that he had been known to kick workmen's tools and clothes out of windows or into excavations in the street if they dared to disagree with him and then flee before their wrath. He also quoted an Irish workman who said that Mylne was a real gentleman, but hot as pepper and proud as a lucifer.[13] Mylne once explained his motives in a letter to James Watt after some engine parts had gone astray: 'I never give up a question of that nature till the party is discovered and has confessed the matter; and that for the good of all concerned, <u>in future</u>.' Put like that, it is just good management.[14]

Another memoir, by John Nichols, seems to hint at this temper, whilst emphasising his undoubted strengths. 'Mr Mylne had peculiarities in his character; but they were chiefly connected with a high independence of spirit, and an inflexible sense of duty and justice. He loved his profession, but not the emoluments of it, and therefore, after all his distinguished employments, did not die rich. Those who knew him could not fail to respect his integrity and admire his talents.'[15]

A historian of the Thames, who had studied Mylne's reports about the locks and weirs of the upper river and noticed his concern for the difficulties of the people who tended them, concluded that he had 'a heart which not all his zeal for improvements nor all his association with stone and iron could rob of its attractive touches of sympathy.'[16]

Mylne showed sympathy for those who fought oppression, together with a wish for the free communication of scientific discoveries. He had known Benjamin Franklin when he was in London, and was said to have been a strenuous advocate of his when it became the fashion to decry him after his attempts to repeal the Stamp Act, which Franklin had believed would avoid the otherwise inevitable war for American independence. Years later, when Britain had effectively lost that war and Mylne's friend Caleb Whitefoord was in Paris as secretary to the British delegation seeking to make peace with the colonists represented by Franklin, they exchanged regards, and Mylne sent the wish that there would soon be peace so that 'the ingenious of all Countrys will enjoy themselves in a free communication of sentiments together, as formerly.' Unusually for an engineer he was a member of the council of the Royal Society at that time, when it voted to send copies of their proceedings to the new college at Harvard despite the war.[17]

Among architectural historians Christopher Hussey thought Mylne 'combined engineering with particularly sensitive neo-classical country houses ... Tusmore is a great neo-classical house *à la française*...' and called The Wick 'one of the most perfect of small Georgian houses'.[18]

Colvin thought that the exteriors of his houses were characterised by a fastidious restraint prophetic of the neo-classical simplicity of the 1790s, and that he was sometimes strikingly original.[19]

Other scholars have continued the note of approval. Ida Darlington thought that his work on the approaches to Blackfriars showed him to have been 'a singularly upright and conscientious public servant.'[20] Charles Hadfield thought him possibly the century's greatest expert on the engineering problems of river navigation.[21]

Another and very different side of his character was expressed in a fondness for tree-planting and gardening, which he did almost everywhere he lived. It is noted at Croydon, Greenwich, New River Head, Amwell and Powderhall. Soon after moving to Islington he obtained permission to make a garden around the upper reservoir. At Amwell he delighted in the productive walnut trees, noting the bushels he had been able to harvest for family use and for sale despite the depredations of bargees from the nearby River Lea navigation, and one of his last actions was the purchase of land with more walnut trees in November 1810. The family usually kept livestock – horses, a pair of goats, always one or two cows for milk, while at New River Head he rented fields to feed them and fatten their calves for sale, and later did the same on a larger scale at Amwell.[22]

At the time of his death Mylne's work was everywhere. Blackfriars with its broad and elegant approach roads was still the newest and most central of London's bridges. Elsewhere were bridges, hospitals, a church, public offices, a concert hall, assembly rooms, an anatomy theatre and museum, castle interiors, country houses large and small, obelisks, and mausoleums. He had embanked a lengthy section of the Thames and designed systems of locks and weirs to make the upper reaches properly navigable. He had improved the New River by replacing leaky aqueducts with proper embankments and adding a waterwheel of his own design, as well as introducing steam engines and the first of its iron mains. At

St Paul's he had cherished the building, carried out works to strengthen the dome and provided a new pulpit, not to mention his work both known and unknown on Nelson's tomb. He had laid out the line of at least one canal and advised on many others.

Mylne left at Inveraray in 1774 a design for what would have been the world's first iron bridge, a design for Inveraray town bridge in the Chinese taste, but it was never built.[23]

Sadly very little remains. The obelisk in St George's Circus and Myddelton's monument at Amwell are still in place, as is the family mausoleum there. A few good houses like The Wick at Richmond (illustration 32) and Pitlour in Fife,[23a] seem little altered, but most of his work has been demolished or extended beyond recognition as needs and fashions have changed. Even bridges such as Romsey that look original turn out to have been widened and reconstructed for heavier traffic, though the 'Atlantic Bridge' at Seil in Argyllshire may be much as built. The largest concentration of his work is to be seen at Inveraray, but the loss of the church spire there does not help its appearance.

Of his social life we know little because his diaries ignore it. This is frustrating because as well as his professional contacts he knew the leading figures of his day through the Royal Society and the clubs that he attended. A record of such meetings spanning half a century would be fascinating. Only from other sources do we learn, for example, of his fondness for wine, conversation and good food, or that guests at his table included Johnson and Boswell. We are unlikely to learn much more about that side of his life, yet it is important not to forget that it was there, alongside his family life and working commitments.

It is also unfortunate that, apart from a couple of mentions by the German archaeologist Winckelmann in the 1750s, no trace remains of his research into the antiquities of Sicily and his subsequent writings on that subject. Perhaps they are somewhere waiting to be discovered.

He was one of the last to excel at both architecture and engineering, and he helped to put both professions on a proper footing. As a memorialist he was successful in having Smeaton commemorated in print, Paterson in oils, and Myddelton, Wren and Nelson in stone. But his own memorial has never been forthcoming, and he probably never foresaw how soon the pace of urban change, railway development and wartime bombing would sweep so much of his work away. His son William never published his papers, and though a grandson began a history of the Mylnes later completed and published by his son, it covers five centuries and gives only a chapter to Robert's life.[24] Although the family carefully preserved most of his papers for more than 150 years, before generously presenting them to archives in London and Edinburgh where their future is secure, Mylne's folio volume of architectural drawings passed out of the family and over two hundred were later sold at Christie's in 1983. At that sale the elevations and plans for Tusmore made £4,500, but most lots sold for less than a thousand pounds.

As a result they are now widely dispersed, most to museums and institutions in Canada and the USA.[25]

Reading the reports he wrote it is clear that he put public benefit above private profit, repeatedly stressing how the public would gain from this work or that and expressing his contempt of the mean private interests that always block attempts to improve cities and waterways. He tried to use his talents for the public good, and saw his works in their wider social context, as can be seen in his years of work on the embankments and approach roads at Blackfriars, the navigation of the upper Thames and the unpaid work he did for bodies like the British Fisheries Society. He had written in 1770 'money I always despised', and over forty years later his obituarist confirmed this by noting how he had loved his profession but not its emoluments.

For all these reasons he deserves to be better known. Proud of his descent from a line of master-masons to Scottish kings, he combined practicality, determination and artistic imagination in a rare blend, while his fierce defence of his honour against the attacks that characterised eighteenth-century England shows how unyielding his integrity was. It probably helped him to keep going after his devastating family losses in the 1790s, for as he had written to his brother in 1777 'this life is chequered, but we must make the most on't. We have no choice and it is useless to repine.' When he planned his own secret tribute to Nelson he aligned himself with the man now seen by many as England's greatest hero, but who he feared might be forgotten, and in doing so demonstrated by the discretion of the deposit that his actions were done not for public show, but because he thought them right. Now he lies nearby, his grave fittingly placed between Wren's and Nelson's, a situation that would have pleased him if he knew of it as his life drew to a close.

Sources

Manuscript

The Mylne family papers have been deposited partly at the British Architectural Library, including Mylne's diaries, many letters and a few drawings, and partly at the National Archives of Scotland. As will be seen from the notes below other relevant papers are to be found in many other archives and libraries.

Robert Mylne's own writings

'Publicus' (pseudonym, but generally accepted to be Mylne), *Observations on bridge building, and the several plans offered for a new bridge across the Thames at or near Black Friars*, 1760. (Copies in BL, Guildhall, Bodleian)

Several of Mylne's reports on Blackfriars, London Bridge and the Port of London are reprinted in vol.132 of Lambert (ed.), *House of Commons sessional papers of the eighteenth century*, Wilmington, 1975

Mylne's principal reports include:

Report to Common Council on London Bridge Water Works, 1767

Report relative to Tyne Bridge at Newcastle, 1772

Report on Wells Harbour, 1781

Report to Common Council on Blackfriars Bridge, Embankment and Surrey Roads, 1784 (in PP vol.xiv, 1799 and 1803, sel cttee App E.1)

Final Report to Common Council on building and completing Blackfriars Bridge, 1786, (in PP vol.xiv, 1799 and 1803 sel cttee App E.2)

Mr Mylne's report on the navigation from Lechlade to Abingdon, 1791

Report of Robert Mylne, Engineer, on the proposed improvement of the drainage and navigation of the River Ouze by executing a straight cut, from Eau-Brink to King's-Lynn, 1792 (but dated 26 October 1791)

Report to the commissioners for improving the navigation of the Thames and Isis from Staines ... to Cricklade, 8 May 1791, and further report on section from Abingdon to Whitchurch, 10 August 1791

Report on a Survey of the River Thames from Boulter's Lock to the City stone near Staines, and on the best method of improving the Navigation of the said River, and making it into as compleat a state of perfection as it is capable of, 24 August 1793. (A copy at LMA Misc Mss 36/20)

Reports on the River Thames, London Bridge and the navigation of the river, 30 May 1800, 15 May

and 30 Oct 1801. (In PP vol.xiv, 1799 and 1803, Sel Cttee App A.1, B.2.)
Scheme or outline for a new London Bridge and improvements to the Port of London, 23 June 1800, (in PP vol.xiv, 1799 and 1803, Sel Cttee App B.2)
Report on three new cuts for the improvement of the navigation of the River Thames above Oxford, 1802

Main published sources

Colvin and Skempton contain, respectively, reliable lists of Mylne's main architectural and engineering works.

Baldwin, R., *Plans, elevations and sections of the machines and centering used in erecting Black-Friars bridge*, 1787 (GH)
Barney, John, *The trials of Wells Harbour*, 2000
Bell, R.F., *Memorials of John Murray of Broughton, sometime secretary to Prince Charles Edward 1740-1747*, 1895
Bynum, W.F. and Porter, Roy, eds, *William Hunter and the eighteenth-century medical world*, 1985
Colvin, H.M., *Biographical dictionary of British Architects*, 1994
Colvin, H.M., ed., *History of the Kings Works*, vol. vi, 1782–1851, 1973
Conway-Jones, Hugh, *The Gloucester and Sharpness Canal*, 1999
Davis, Bertram Hylton, *Johnson before Boswell, a study of Sir John Hawkins life of Samuel Johnson*, 1961
Dennistoun, James, *Letters of Sir Robert Strange Knight, engraver … and of his brother-in-law Andrew Lumisden &c*, 1855
Dickinson, H.W., *Matthew Boulton*, 1936, reprinted 1999
Doty, Richard G., *The World Coin, Matthew Boulton and his industrialisation of coinage*, Interdisciplinary Science Reviews, vol.15, No.2, 1990, pp 177–186
Doty, Richard G., *The Soho Mint and the industrialisation of money*, 1998
Downes, K., *Sir Christopher Wren: the designs for St Paul's Cathedral*, 1988
Dugdale, Sir William, *The history of St Paul's Cathedral in London, from its earliest foundation*, 1818
Dunlop, Jean, *The British Fisheries Society, 1786-1893*, 1978
Elmes, James, *Memoirs of the life and works of Sir Christopher Wren*, 1823
Elmes, J., 'A history of architecture in Great Britain', *The Civil Engineer & Architect's Journal*, 1847, vol.10, p.340
Farington, J., ed. Cave, *Diary of Joseph Farington*, 1983
Fleming, John, *Robert Adam and his circle*, 1962
Golan, Tal, *Laws of man and laws of nature*, Harvard, 2004
Gifford, John ed., *Buildings of Scotland – Edinburgh*, 1984
Gotch, Christopher, *The Gloucester and Sharpness Canal and Robert Mylne*, 1993
Grosley, Pierre Jean, *New observations on Italy and its inhabitants*, 1769
Grosley, Pierre Jean, *A tour to London, or new observations on England and its inhabitants*, 1772
Hadfield, Charles, *Hadfield's British canals*, 1998
Hadfield, E.C.R., *The canals of Southern England*, 1955
Harris, Eileen, *British architectural books and writers, 1556–1785*, 1990
Houfe, Simon, *Sir Albert Richardson, the professor*, 1980
Ingamells, John, *A dictionary of British and Irish travellers in Italy, 1701–1800*, 1997
Keene, Derek, Burns, A., Saint, A. (eds), *St Paul's, the cathedral church of London, 604–2004*, 2004
Lindsay, I.G. and Cosh Mary, *Inveraray and the Dukes of Argyll*, 1973
Mason, Shena, *The Hardware Man's Daughter*, 2005
Mylne, R.S., *The master masons to the crown of Scotland and their works*, 1893

Richardson, A.E., *Robert Mylne, Architect and Engineer 1733–1811*, 1955
Richardson, George, *The new Vitruvius Britannicus*, 1802
Rudden, Bernard, *The New River, a legal history*, 1985
Ruddock, Ted, *Arch bridges and their builders 1735–1835*, 1979. This gives a full and detailed account of the building of Blackfriars Bridge
Ruddock, Ted, *Travels in the colonies in 1773–1775 described in the letters of William Mylne*, Atlanta, Georgia, 1993
Skempton, A.W. ed., *A biographical dictionary of civil engineers in Great Britain and Ireland, vol. 1, 1500–1830*, 2002
Smiles, Samuel, *Lives of Boulton & Watt*, 1865
Smeaton, John, *Reports of the late John Smeaton*, 1797 and 1812
Stillman, Damie, *English Neo-Classical Architecture*, 1988
Stillman, Damie, British architects and Italian architectural competitions, *Journal of Society of Architectural Historians*, Philadelphia, vol.32, 1973
Stroud, Dorothy, *George Dance Architect 1741–1825*, 1971
Ward, Robert, *London's New River*, 2003
Watkin, David, *Athenian Stuart: Pioneer of the Greek revival*, 1982
Winckelmann, J. J., ed. Gross et al, *Schriften zur antiken Baukunst*, Leipzig, 2001
Woodley, R.J. *Robert Mylne (1733–1811) The bridge between architecture and engineering*, University of London, unpublished doctoral thesis containing a very full and careful account of Mylne's life and work, 1998
Woodley, R. 'Professionals: early episodes among architects and engineers', *Construction History*, vol.15, 1999
Wren, C. *Parentalia, or Memoirs of the Family of Wrens*, 1750

Greenwich Hospital and other naval matters

The case of the Royal Hospital for Seamen at Greenwich, 1778
State of facts relative to Greenwich Hospital, 1779
Another state of facts relative to Greenwich Hospital, 1779
A solemn appeal to the public from an injured officer, Captain Baillie ... and the evidence given on the subsequent enquiry at the bar of the House of Lords, 1779
Anon., *The Case and Memoirs of the late Rev. Mr James Hackman and of his acquaintance with the late Miss Martha Reay*, 1779
The Speech of the Earl of Sandwich in the House of Lords on Friday the 14th day of May 1779, Being the fourteenth day of the sitting of the Committee of Enquiry into the Management of Greenwich Hospital. Printed for Cadell, T., opposite Catherine Street, Strand, 1779
Bold, John, *Greenwich: an architectural history of the Royal Hospital for Seamen and the Queen's House*, 2000
Cooke, John and John Maule, *An historical account of the Royal Hospital for Seamen at Greenwich*, 1789
Czisnik, Marianne, *Horatio Nelson, a controversial hero*, 2005
Mann, A.Y., *The last moments and principal events relative to the ever to be lamented death of Lord Viscount Nelson with the procession by water and the whole ceremony of the funeral, intended as a sequel to his life*, 1806
Nicolas, Sir N.H., *The dispatches and letters of Vice Admiral Lord Viscount Nelson*, 7 vols, 1844–46
Pocock, Tom, *The terror before Trafalgar*, 2002
Rodger, N.A.M., *The Wooden World, an anatomy of the Georgian navy*, 1986
Rodger, N.A.M., *The Insatiable Earl – A life of John Montagu, fourth Earl of Sandwich, 1718–1792*, 1993
Rodger, N.A.M., *The Command of the Ocean, a Naval History of Britain 1649–1815*, 2004

Notes

Abbreviations

BAL	British Architectural Library
BCA	Birmingham City Archives (Archives of Soho)
BL	British Library
BM P&D	British Museum, Department of Prints and Drawings
CC	Common Council of the City of London
CLRO	City of London Record Office, mostly now at LMA
ESRO	East Sussex Record Office
GH	Guildhall
GM	*Gentleman's Magazine*
HMC	Historical Manuscripts Commission
ICE	Institution of Civil Engineers
ILN	*Illustrated London News*
LMA	London Metropolitan Archives
LPL	Lambeth Palace Library
NAS	National Archives of Scotland
NLS	National Library of Scotland
NMM	National Maritime Museum
Notts	Nottingham University Library
ODNB	Oxford Dictionary of National Biography
PP	Parliamentary Papers
RA	Royal Academy
RM	Robert Mylne
TM	Thomas Mylne
TNA	The National Archive, formerly Public Records Office
ULSH	University of London, Senate House
WM	William Mylne

Names with dates or no other details refer to authors listed in *Sources* above.

Chapter 1 Edinburgh pages 11–17

1 McAdam, A.D. and Clarkson, E.N.K., *Lothian Geology*, 1996
2 Mackenzie, Henry, author of *The Man of Feeling*, in an unpublished work cited by

William Steven in *History of the High School of Edinburgh*, 1849, p.102. See also Trotter, J.J., *The Royal High School, Edinburgh*, 1911
3 McKean, *Architecture of Robert Adam*, 1992, p.41
4 BAL MyFam 1/1; NAS GD 1/51/44 & letters home
5 Gifford, John ed., *Buildings of Scotland – Edinburgh*, 1984; and see R.M. Pinkerton and W.J. Windram, *Mylne's Court*, 1983
6 NAS GD 1/51/37; Gifford pp.162–3
7 Trotter, op. cit. p.41, and Daiches, D., *Charles Edward Stuart*, 1973
8 James Buchan, *The capital of the mind*, 2003; Campbell, John, *The diary of John Campbell – a Scottish banker and the '45*, 1995
9 BAL MyFam 2/12/1-2
10 Atholl, John, 7th Duke, *Chronicles of the Atholl and Tullibardine Families*, 1908, vol. 3, pp.467–68

Chapter 2 Paris pages 19–23

1 BAL MyFam/4; NAS GD 1/51
2 Fleming, p.188
3 *The gentleman's guide in his tour through France, wrote by an officer in the Royal-Navy*, 1766
4 BAL MyFam 4/13, 3 January 1755. RM to TM

Chapter 3 Rome pages 25–34

1 But see e.g. BAL MyFam 4/16, 11 February 1756, RM to TM
2 Dennistoun, James, *Letters of Sir Robert Strange Knight ... and of his brother-in-law Andrew Lumisden*, 1855, Letters, to Richard Morrison, 2 November 1756, and to his sister Lady Strange, 7 October 1758
3 BAL MyFam 4/14, 12 May 1755, Burnet to WM
4 BAL MyFam 4/15, 24 September 1755 RM to TM
5 BAL MyFam 4/16, 11 February 1756
6 BAL MyFam 4/19, 4 January 1757, RM to TM
7 BAL MyFam 4/16; 11 February 1756
8 BAL MyFam 4/17, 30 June 1756
9 BAL MyFam 4/19, 4 January 1757, RM to TM
10 Fleming, p.188
11 Fleming, p.212
12 BAL MyFam 4/16, 11 February 1756, RM to TM; Clerk of Penicuik MSS, GD 18/4807, cited in ODNB
13 BAL MyFam 4/23, December 4 1757
14 BAL MyFam 4/28, 16 February 1758, WM to RM
15 BAL MyFam 4/20, 6 August 1757, WM to TM
16 BAL MyFam 4/21, 4 October 1757, WM to TM
17 BAL MyFam 4/22, 17 November 1757, WM to RM
18 BAL MyFam 4/24, 7 December 1757, WM to RM; 4/27, 28 January 1758, RM to WM
19 BAL MyFam 4/23, 4 December 1757 WM to TM; 4/25, 13 January 1758, RM to TM; 4/26, 13 January 1758, WM to RM
20 BAL MyFam 4/28, 16 February 1758, WM to RM
21 BAL MyFam 4/31, 26 April 1758, RM to WM
22 BAL MyFam 4/32, 1 June 1758, WM to RM

23 BAL MyFam 4/33, 31 [sic] June 1758, RM to WM
24 BAL MyFam 9, Printed copy of letter to Professor Robison in Surveyor's Private Book, 1799
25 BAL MyFam 4/31, 26 April 1758, RM to WM
26 Ingamells, citing William Patoun *c.*1766
27 Laurie & Whittle, 1799
28 BAL MyFam 4/27, 28 January 1758, RM to TM
29 BAL MyFam 4/34, 23 September 1758

Chapter 4 '... thump, thump, thump, I feel it yet...' pages 35–40

1 BAL MyFam 4/34, 23 September 1758, RM to WM
2 *Delle lodi delle belle arti &c,* publ. Pagliarini, Rome, 1758, copy in NLS; Grosley, translated by Nugent, *New observations on Italy and its inhabitants*, 1769, vol.2, p.161–4
3 *Edinburgh Evening Courant,* 21 October 1758
4 BAL MyFam 4/37, 16 December 1758, RM to WM
5 Damie Stillman, 'British Architects and Italian architectural competitions', *Journal of the Society of Architectural Historians, Philadelphia*, vol.32, 1973
6 Archives, Accademia Parmense di Belle Arti, Parma; cited in Stillman, Damie, 'British Architects and Italian architectural competitions', see above
7 BAL MyFam 4/38, 18 January 1759
8 BAL MyFam 4/39, 27 February 1759
9 NAS, Clerk of Penicuik MSS, GD 18/4889, 7 March 1761
10 HMC 28, MSS of the Earl of Charlemont, vol.1, p.252, 1 April 1759
11 BAL MyFam 4/40, 6 April 1759
12 BAL MyFam 4/41, 10 June 1759, and see Ingamells
13 BAL MyFam 4/42, 24 June 1759
14 BAL MyFam 4/43, 10 July 1759

Chapter 5 London pages 41–56

1 BAL MyFam 4/34, 23 September 1758, RM to WM
2 BAL MyFam 4/44, 17 July 1759, RM to WM
3 BAL MyFam 4/44, 17 July 1759, RM to WM
4 BAL MyFam 4/46, 18(?) August 1759, RM to WM
5 City of London Records, Jo. 45, 423; *Debates of the House of Commons from the Year 1667 to 1694*, collected by Hon. A. Grey, 1769; both cited in Smiles, *Lives of the Engineers – Harbours, Lighthouses, Bridges*, 1874, pp.67–68
6 BAL MyFam 4/31, 26 April 1758, RM to WM
7 BAL MyFam 4/37, 16 December 1758, RM to WM
8 BAL MyFam 4/46, 18(?) August 1758, RM to WM
9 BAL MyFam 4/47, 25 August 1759
10 BAL MyFam 4/48, 30 August 1759
11 Bridge Committee Report in Journal of CC 18 January 1771, CLRO. Also in Guildhall Library A8/5–35, cited in Ruddock, p.63
12 BAL MyFam 4/49, 24 January 1769
13 e.g. BL, Guildhall, Bodleian, Eighteenth century collections online
14 BAL MyFam 4/48, 30 August 1759, RM to WM
15 Damie Stillman, *English Neo-classical Architecture*, 1988

Chapter 6 '… against the caballing interest…' pages 57–77

1 BAL MyFam 4/50, 23 February 1760, RM to TM
2 BAL MyFam 4/49, 24 January 1760; RM to TM
3 BAL MyFam 4/53 London May 15th 1760 RM to Dear Father
3a See BM P&D, Satires 3733, 3734 and 3741, and a variant at BAL MyFam/9. Guildhall Library has a copy of the ballad *Bute-Awl*.
4 John Gwynn, *London and Westminster Improved*, 1766, p.64
5 'Argument for and against Elliptical Arches and Iron Rails' in *London Chronicle*, 1–4 December 1759, unsigned letter, presumably from Mylne
6 *Publicus*, p.24–5
7 *Publicus*, p.31
8 M. Grosley, translated by T. Nugent, *A Tour to London, or observations on England and its inhabitants*, vol.1, 1772
9 Select Committee on the Port of London, Third Report 1800, App. A1, 30 May 1800
10 TNA CRES 38/1210, and see diary 25 March 1776
11 BAL MyFam 4/54 London 13 September 1760, RM to WM
12 BAL MyFam/9
13 See entries for publications by Ruddock and Woodley in 'Sources' above
14 BAL MyFam/9, 14 April 1760
15 Select Committee on the Port of London, Third Report 1800, App to E.1, paras xiii, xiv, xxxviii, xl–xlii
16 BAL MyFam 4/52, 13 March 1760, TM to RM
17 TNA, T 29/34, 24 June 1761, Treasury minute book, re 'Petition of Messrs Wal-ger and Fletcher for leave to dig Portland Stone for the building Black Fryars Bridge'
18 TNA T1/414. f.84, 16 January 1762, and see TNA T 29/34, 21 January 1762
19 Boswell, James *Life of Johnson*, 1953 edition, p.248 n.2
20 BAL MyFam 4/53 15 May 1760, TM to RM and 4/57, 15 November 1760, RM to TM
21 Warwickshire County Record Office, L6/1464, 21 March 1762, letter George Lucy to another
22 BAL MyFam /9, f.111; GM vol. 36, June 1766
23 Walker and Burges, 'Report on Blackfriars Bridge', 2 March 1833, in *Report to Common Council*, 25 April 1833
24 Thornbury, Walter, *Old and New London*, vol. 1, *c.*1880
25 Walker and Burges, 'Report on Blackfriars Bridge', 2 March 1833, in *Report to Common Council 25 April 1833*
26 BAL MyFam /9, 14 April 1760
27 BAL MyFam/9, 20 May 1760
28 BAL MyFam/9 26 August 1760
29 BAL MyFam/9 f253
30 William Falconer, *An universal dictionary of the marine*, 1769, 'Pump'
31 GM 1761, vol. 31, p.236, and see BM P&D Satire 3734
31a See Winckelmann, J.J., *Anmerkungen über die Baukunst der alten Tempel zu Girgenti in Sicilien*, 1762
32 Printed copy of letter to Professor Robison in BAL MyFam/9
33 H. Chamberlain, *History and survey of the cities of London and Westminster &c*, 1770, p.375
34 *Annual Register* 1761, p.124
35 *Annual Register* 1770, p.176; Lecture VI, cited in Watkin, David, *Sir John Soane, Enlightenment thought and the Royal Academy lectures*, 1996; John Summerson '*Georgian London*', 1945
36 Smiles p.71

Chapter 7 '… on the pinnacle of slippery fortune…' pages 79–95

1a BAL MyFam 4/57, 15 November 1760 RM to TM.
1b BAL MyFam 4/36, 1758, WM in Edinburgh to RM in Rome and MyFam 4/53, London 15 May 1760 RM to TM
2 BAL MyFam 4/33, 'June 31' 1758, RM to WM
3 BAL MyFam 5/1, 24 February 1763
4 BAL MyFam/12, 13 and /14
5 17 July 1762
6 A.E. Richardson, *Robert Mylne, architect and engineer 1733–1811*, 1955; Colvin 1994 ed.
7 ODNB, James Boswell
8 *Boswell's London Journal 1762–1763*, ed. F.A. Pottle, 1950, entries for 23 and 29 November 1762
9 NAS GD 10/1421, vol. 6, f.287, 26 January 1759, RM to Lord Garlies
10 Survey of London, vol. 31 p.48; NAS GD/113/5/209/ Bundle 1
11 NAS RHP 8822, 26 January 1759, RM to Lord Garlies
12 Notts PWF 7100, 1 September 1766 RM to Duke of Portland
13 Macauley & Greaves, eds. *The autobiography of Thomas Secker, Archbishop of Canterbury*, University of Kansas, 1988
14 Farington, *Diary* 6 May 1811; Elmes, James, *A General and Bibliographical Dictionary of the Fine Arts*, 1826, unpaginated, entry 'Architecture'
15 1764–6, NAS GD1/51/46(2), and see Notts PwF 7046/1-2; 7047; 7094-5
16 *The Thames at Blackfriars c.1750*, school of Samuel Scott, Guildhall Art Gallery
17 BM P&D 1862-12-13-51
18 Patent 395 of 1714
19 BAL MyFam 5/5, 13 September 1770, RM to Elizabeth Duncan
20 D.G.C. Allan & John L. Abbott, *The Virtuoso Tribe of Arts and Sciences: Studies in the Eighteenth-century work and membership of the London Society of Arts*, 1992
21 Annual Register 1771
22 Bentham, J., ed. M.J. Smith & W.H. Burston, *Chrestomathia*, 1983
23 Diary, 1775
24 Edgeworth, Richard Lovell, *Memoirs*, 1820. vol.1, pp.188–9, 381
25 14 May 1773, RM to Hamilton, BL Add. 42069 ff 89 & 125
26 BAL, MyFam 1/1
27 A.J. Youngson, *The Making of Classical Edinburgh 1750-1840*, Edinburgh, 1988
28 Lady Maxwell's journal, quoted in Rev. John Lancaster, *The life of Darcy, Lady Maxwell, late of Edinburgh*, 1826, p.48
29 *Reports of John Smeaton*, 1812, vol.3, reports of 22 August 1769 and 27 January 1770

Chapter 8 '… this best step of my life…' pages 97–122

1 BAL MyFam 5/5, 13 September 1770, RM to his mother Elizabeth Duncan
2 6 October 1770. NAS RH/15/44/33
3 Grosley 1772, p.44, 46, 294
4 London, 8 July 1771, John Gray to Tobias Smollett
5 BL Mss, Add 36593, f.172
6 Edinburgh GD1/51/99, 19 September 1777, RM to WM
7 *Reports of the late John Smeaton*, vol.1, 1797, Preface
8 *Scots Magazine*, October 1772, cited in Lindsay & Cosh
9 BAL, MyFam 11/3/4, burgess tickets 9 and 10 September 1772

10 *Book of the Old Edinburgh Club*, vol.ix, 209
11 BAL MyFam 4/, 29 August and 4 September 1773, WM to AM
12 BAL MyFam 4/, 7 March 1774, copy by WM of letter RM to WM, in letter 13 October 1774 WM to AM
13 A very full and annotated account of William's travels, with transcripts of the correspondence, mostly between him and Anne, is in Ted Ruddock, *Travels in the colonies in 1773–1775 described in the letters of William Mylne*. Atlanta, Georgia, 1993
14 Boswell, *Journal*, September 1774
15 NAS GD 1/51/98/3
16 *Country Life*, 1 February 1941
17 Nicholas Goodison, *Matthew Boulton – Ormolu*, 2002; Eric Robinson & E. Musson, *James Watt and the steam revolution* 1969; Margaret C. Jacob, *Scientific culture and the making of the industrial west*, 1997; BCA MS 3147 MS 4/53
18 LMA, H10/CLM/A1/2, Minutes 1767–78
19 For a fuller account of the delay in payment see Roger Woodley, 'A very mortifying situation: Robert Mylne's struggle to get paid for Blackfriars Bridge', *Architectural History* 43:2000, 172–186
20 Notts PwF 7104, RM to Portland, 13 October 1767
21 Reynolds' fee for the portrait was 35 guineas, with a further £26 to Thomas Watson for a mezzotint engraving. The two dinners cost £66 – Diary, 1776
22 Grosley 1772, vol.2 p.93
23 LPL, MS Lowth 1, ff 195–197
24 LPL MS Cornwallis 3, ff 76–81
25 LPL MS 1489, f.63
26 Ward, *London's New River*, pp.154–6
27 Diary, February 1770
28 TNA ADM/67/11; NMM San/F/37a/64 and /36
29 Grosley, 1772
30 John Cooke and John Maule, *An historical account of the Royal Hospital for Seamen at Greenwich*, 1789
31 9 October 1776, RM to board, TNA, ADM 65/106
32 23 December 1777, Allwright to Sandwich, NMM San/F/37a/35
33 9 April 1777, Baillie to Sandwich, NMM San/F/37a/33
34 George Adolphus, *History of England from the accession of George III to the conclusion of the peace in the year 1783*, 1802, vol.1
35 TNA, ADM 65/106, 20 April 1777
36 NMM, SAN/F/37a/36 – 1 April 1778, RM to Sandwich
37 *Bye-laws, rules, orders and directions for the better government of His Majesty's Royal Hospital for Seamen at Greenwich*, Admiralty, 16 February 1776, 'Clerk of the Works, Art. 3'. BL 797.dd.7
38 *True copies of affidavits filed in the Court of King's Bench &c*, 1779, NMM 094 E 4654; *State of Facts Relative to Greenwich Hospital, 1779*, TNA ADM 80/93
39 TNA ADM 65/106
40 Anonymous pamphlet, *Another state of the facts relative to Greenwich Hospital*, March 1779, in TNA ADM 80/92, and BL
41 Anon. *The Case and Memoirs of the late Rev. Mr James Hackman and of his acquaintance with the late Miss Martha Reay*, 1779; N.A.M. Rodger, *The Insatiable Earl – A life of John Montagu, fourth Earl of Sandwich, 1718–1792*, 1993; *Walpole Correspondence*, ed. Lewis, vol.33 pp.99–100
42 NMM Caird Library, *The Speech of the Earl of Sandwich in the House of Lords on Friday the 14th day of May 1779, Being the fourteenth day of the sitting of the Committee of*

Enquiry into the Management of Greenwich Hospital. Printed for T. Cadell, opposite Catherine Street, Strand, 1779

43 Parliamentary Register vol.14, 1778–9, p.301, and see TNA ADM 67/145, p.147; and *A solemn appeal to the public from an injured officer, Captain Baillie … and the evidence given on the subsequent enquiry at the bar of the House of Lords,* 1779

44 4 February, 26 September, 20 December 1779, RM to Sandwich, NMM San/F/37a/41/53/54

Chapter 9 '… a man of pure honour…' pages 123–150

1 LMA ACC 2558, NR/1, Minutes July 1780, 17 August, 19 October, 21 December 1780

2 Adolphus, G. *History of England from the accession of George III … to 1783,* 1802, vol. 1 p.293

3 *Particulars of Anglesey mines,* BL Add. 38375, f. 96; NAS 2362, bundle 156, 10 February 1779

4 Diary

5 NAS GD 1/51/37, *Notes of Journey into Scotland A'o 1781*

6 Lindsay & Cosh; Lot 31, Christie's sale 30 November 1983; NAS GD 1/51/37; Diary passim

7 Letters 3 Oct 1775 AM to WM, 7 March '1744' actually 1774 WM to AM

8 V.C.P. Hodson, *List of the officers of the Bengal Army, 1758-1834,* 1928

9 BCA, MS 3219/3/396/56, RM to MB 24 October 1787, and on this point generally see N.A.M. Rodger, *The Command of the Ocean, a Naval History of Britain 1649–1815,* 2004, p.420

10 TNA ADM 67/30, ff 29, 180, 188, 192, various dates March–December 1781

11 TNA ADM 65/106

12 TNA, ADM 67/12, p.10

13 LMA ACC 2558/MW/C/15/243/3

14 NMM San/F/37a, /56 and /61

15 TNA, Original letter in ADM 65/106; minutes in ADM 67/31

16 BCA MS 3782/12/93/16, Montagu to Boulton 10 August 1780, cited in Mason 2005; Montagu to Leonard Smelt, 26 April 1780, Huntingdon Library MO 50256 cited in David Watkin, *Athenian Stuart: Pioneer of the Greek revival,* 1982, p.50; TNA ADM 65/106, 22 February 1783, Affidavit of Dominick Bartoli

17 RM to Sandwich, 6 April 1783, NMM San/F/37a/64

18 TNA C12/595/11, C12/615/27; 4 July 1781, TNA ADM 67/12 p.10; 12 May 1799; Printed copy of letter to Professor Robison in BAL MyFam/9

19 BL shelfmark 1607/3465 – Second report of the Committee of Common Council, presented 25 November 1784

20 Sutro Library, California, MS Banks H 1:75, *Mr Serjeant Brown's Minutes of a Conference betwixt Mr Milne, Mr Elstobb and himself,* 14 May 1781

21 *Folkes v. Chadd,* (1782) 3 Doug. KB 157

22 University of London, Senate House MS 844

23 ICE SM/ML/1, p.180 Smeaton to Mr Forster, 1 July 1783

24 For a full analysis of the complex legal issues raised in the course of the Wells Harbour litigation see Tal Golan, *Laws of man and laws of nature,* Harvard, 2004, Chapter 1

25 *Correspondence of Horace Walpole,* ed. Lewis, vol. 35, 1973, pp.633–4

26 Isaac Gumley, 'Wells Triumphant' in *Poetical essays on various subjects,* Market Harborough, 1797

27 ICE SM/ML/2/24-25, 4 November 1783, Smeaton to Forster

28 John Barney, *The trials of Wells Harbour,* 2000; Golan, op. cit. p.47

29 Robert Baldick, 'The Duel', 1965

30 MP and Lord Clerk Register from 1756 – Hogarth's print 'Paul before Felix' is said to show him as Tertullus
31 Lewis M. Knapp, ed., *The Letters of Tobias Smollett*, Oxford, 1970, Letter of 25 February 1753, pp.21–26
32 Lt Samuel Stanton of the 97th Regiment, *The Principles of Duelling*, 1790
33 Richard Hey LLD, *A short treatise upon the propriety and necessity of duelling*, 1784, BL 526.k.29, and see Blackstone's Commentaries, Bk 4, Ch.2, p145, p150
34 Norfolk Record Office MF/RO 504/2, MS 486, pp.133–8; Holkham Hall; Bodleian, Gough, Norfolk 1(4)
35 ICE, SM/ML/2 p.68, Smeaton to Hardinge, 6 September 1784
36 *Annual Register* 1804, pp.470–473
37 Frederick Robinson to Lord Grantham, 22 May 1778, Bedfordshire and Luton Archives, L30/14/333/97
38 Farington, *Diary*, 27 April 1816
39 George Dempster, *Discourse ... to the British Society for Extending the Fisheries and Improving the Sea Coasts, with Reflections ... by John Gray ...*, 1789. See also James Fergusson (ed.) *Letters of George Dempster to Sir Adam Fergusson 1756–1813*, 1934; and Jean Dunlop, *The British Fisheries Society 1786–1893*, Edinburgh, 1978
40 TNA, ADM 65/106, 12 August 1785, RM to Newton.

Chapter 10 '... sufficiently disagreeable sensations...' pages 151–166

1 LMA, ACC 2558 NR /1/2, f.153, 1782, and NR/1/3, f.141, 1789
2 Pottle, Marion S. et al. *Catalogue of the papers of James Boswell at Yale University*, (1993) vol.1, p.64
3 LMA, ACC 2558, NR 1/3, pp.74, 76, 91, 94
4 ESRO, Lewes; ACC 6420, RM to Mrs Selby, 13 August 1788; BCA, MS 3782/12/33/85, RM to MB
5 BAL, MyFam 2/11/2, Will of Lt William Mylne; NAS GD 1/51/98/11, AG to WM, 8 June 1777
6 BAL MyFam/ 22
7 BCA, MS 3782 12/41/153 and 12/41/392, 27 April and 24 December 1796
8 BAL, Henry Holland's papers, including HoH/2/7; Farington, *Diary*, 2 November 1797, 3 January 1799
9 GM 1791, vol. 61, p.234
10 TNA, T 29/63 f.196, T 1/699 ff 9–10 and 98–9
11 Weaver, C.P. and C.R., *The Gloucester and Sharpness Canal*, c.1967; Conway-Jones, H., *The Gloucester and Sharpness Canal*, 1999; Gotch, Christopher, *Robert Mylne and the Gloucester and Sharpness Canal*, 1993
12 BCA MS 3782 12/34/40
13 Diary, January 1791
14 *Roach's London Pocket Pilot or Stranger's Guide through the Metropolis*, 1793, p.48
15 Thornbury, *Old and new London*, vol.1, p.207
16 Hertfordshire Record Office, architectural drawings for The Grove, Great Amwell. D/EX 440/P1-17
17 BL Add. Ms. 42069, f.125, 10 June 1788, RM to Sir William Hamilton
18 BCA MS 3219/4/39/6, RM to JW, 11 February 1797
19 *GM* vol. 67, p.623
20 BCA MS 3219/4/39/22a, RM to JW, 26 August 1797
21 *Lloyd's List*, 24 October 1797; Diary
22 BCA MS 3147/3/416/94, 10 January 1798

23 TNA, RAIL/829/1, General assembly of Gloucester & Berkeley Canal Company 1793–1823, p.167
24 BAL MyFam 8/6/5; Diary
25 NMM COC/11, personal diary of Admiral Sir George Cockburn Bt., 1797–1818; J.J. Colledge, Ships *of the Royal Navy*, 2003; David Lyon, *The Sailing Navy List – All the ships of the Royal Navy built, purchased and captured, 1688-1860*, 1993, p.243; James Pack, *The man who burned the White House, Admiral Sir George Cockburn 1772–1853*, 1987, p.9
26 NMM COC/11
27 TNA ADM 52/3224, /7 and /5; NMM, ADM.L./M.292
28 Rodger 2004 p.439
29 East Sussex Record Office, ACC 6420
30 East Sussex Record Office, ACC 6420

Chapter 11 '*Si monumentum requiris…*' pages 167–174

1 *The Builder*, 2 January 1864
2 LMA, ACC 2558, NR 1/8 p.174, 25 January 1816
3 Lindsay & Cosh, p.284–3
4 Lindsay & Cosh, p.270
5 Colin McDonald (ed.) *Third Statistical Account of Scotland – The County of Argyll*, 1961, p.221
6 Select Committee, Third Report on the Port of London, Appendix B.1, 23 June 1800, p.51; answers to Select Committee of Bridge House 15 May and 30 October 1801, p.64
7 The upper stone is currently in the Museum of London, while the lower can be seen in the Museum in Docklands
8 *Historical Chronicle*, 28 July 1770
9 TNA, ADM 65/106, 7 October 1778
10 Lundie, Angus, Duncan mausoleum; Holy Trinity, Fareham, Hants, Thompson memorial; Askham Richard, Yorkshire, memorial to Elizabeth Berry, mother of Walpole's Berry sisters; Clowance, Cornwall, St Aubyn mausoleum
11 Wren, C. *Parentalia, or Memoirs of the Family of Wrens*, 1750, p.347
12 Elmes, James, *Memoirs of the life and works of Sir Christopher Wren*, 1823
13 Cited in Keene et al p.272
14 Farington's Diary 11 July 1804
15 Elmes, op. cit. p.525
16 Matthews, W.R. *St Paul's in War and Peace: 1939–1958*
17 TNA, IR 37/226 Pt IV, 22 June 1956 W. Godfrey Allen to War Damage Commission
18 Diary 1767. Wren drawings now at Guildhall Library; Kerry Downes *Sir Christopher Wren: the designs of St Paul's Cathedral*; LPL MS 2552, f. 63 records the purchase, two of the Thornhill drawings are still at St Paul's: see Keene Burns & Saint op. cit.

Chapter 12 '… the Greatest Seaman…' pages 175–195

1 NMM MS 80/050, 11 December 1805, William Haslewood
2 Tyson to Haslewood, 6, 11 & 12 December 1805, Nelson Museum, Monmouth, E 214, 215, 217; cited in Coleman, *Nelson – the man and the legend*, 2001
3 Archibald Duncan, *A correct narrative of the funeral of Horatio Lord Viscount Nelson*, 1806
4 NMM STW/8, 18 and 21 November 1805, Earl Nelson to Col McMahon

5 18 and 22 February 1801, Nelson to Emma Hamilton, cited in Aspinall, *Correspondence of George IV*, vol.5, pp.281–3
6 NMM STW/8, 19 and 21 November 1805, Duke of Clarence to Earl Nelson and v.v.
7 *Diary*, 23–25 November 1805; BL Add MS 34992 f.68
8 TNA LC 2/40 f.77; Mann (1806) op. cit.
9 Naval Chronicle, vol. xiv, p.505, J. Farington, *Diary*
10 J. Farington, *Diary*, pp.2595-6; BCA MS 3782/12/85/134, 18 November 1796, cited in Mason, 2005; Colvin, *History of the King's Works*, vol.6, pp.53, 69
11 BCA MS 3782 12/50/228, 27 December 1805, RM to MB
12 TNA LC 2/40,ff.10, 12, 13, 16, 17; *Naval Chronicle*, xv, p.231
14 NMM PAR 167c, 10 January 1806, St Vincent to Baird; BL Add MS 29,915, 3 January 1806, cited in Terry Coleman, *Nelson – the man and the legend*, 2001
15 James Harrison, *Life of the Rt Hon. Horatio, Lord Viscount Nelson*, 1806, vol. 2, p.456
16 Survey of London, vol. xxv, *St George's Fields*, 1955, p.44
17 Samuel Smiles, *Lives of Boulton & Watt*, 1865, p.267
18 Mayo, John Horsley, *Medals and Decorations of the British Army and Navy*, 1897, p.176, citing Nicolas vol.7 p.230 and vol.4 p.527
19 *Naval Chronicle*, vol. xv, 1806 January–June p.18
20 Margaret C. Jacob, *Scientific culture and the making of the industrial west*, 1997; and see BCA MS 3147 MS 4/53
21 BCA MS 3782/12/33/85, 21 July 1788; and MS 3782 1/10 4 November 1775; Boulton's diary, 15 January & 18 March 1792; BAL MyFam/12 *Diary* September 1775, January 1793
22 BCA Matthew Boulton correspondence, MS 3782 12/50/226, 24 December 1805
23 TNA LC 2/40 f.4
24 BCA MS 3782 12/50/228, 27 December 1805
25 BCA MS 3782 12/50/233, 31 December 1805
26 BCA MS 3782 12/51/2, 1 January 1806
27 BCA MS 3782 12/51/4, 3 January 1806
27a *Horace Walpole's Correspondence*, ed. Lewis, vol.28 p.393 and vol.24 p.62
28 NMM MS 80/050, 30 December 1805, RM to W. Haslewood; Cuttings labelled 'Obs Jan 5 1806' and 'MP Jan 6' in Miss Banks' Collection of Albums, BL shelfmark LR.301.h.3-11
29 *The Times* 2 January 1806 and Guildhall MS 22/298
30 Archibald Duncan, *A correct narrative &c...Nelson*, 1806
31 Adam Collingwood, *Anecdotes of the late Lord Viscount Nelson*, 1806, pp.135–6; T. Tegg, *An official and circumstantial detail of the grand national obsequies at the public funeral of Britain's darling hero, the immortal Nelson*, pp.23–4
32 e.g. Collingwood p.124. Many of the published accounts of the funeral simply repeat each other. *The Times* of 10 January has a useful account
33 *The Times*, 21 January 1806, p.3
34 e.g. W. Dugdale, *The History of St Paul's Cathedral in London ... with additions by Henry Ellis*, 1818, p.213. TNA T 29/91 13 August 1807, C.Div: No. 1.356; T 27/60 f.133 21 August 1807; T 29/93 2 February 1808, 6th Div. No. 644; T 27/61 f.47 9 February 1808. GH prints 12/34
35 Colvin, *History of the Kings Works* vol.vi pp.53, 69, 97-8; TNA T 29/122 f.526, 6 April 1813; Farington, *Diary* pp.2595-6, 6396-7 and 7389; TNA WORK 4/21, 3 June 1814; ESRO FRE/2826 Henrietta Fry to Mary Frewen, 21 July 1810
36 RA Council minutes 20 June 1806; letter from Richard Westmacott 2 December 1852, RA, AND/26/371; TNA LC 2/40
37 *ILN* 13 November 1852; *The Builder*, 30 October 1852

38 GH Ms 25809, passim; TNA WORK 36/53; *The Builder*, 1 Feb 1879; *The Builder*, 1 June 1956, p.626
39 TNA LC 2/38/2
40 BAL MyFam /9, 'Surveyor's Private Copy', 16 October 1760
41 Westwood, Arthur, *Matthew Boulton's medal on the reconquest of Naples in 1799 &c*, 1926

Chapter 13 '... mine has been a very active life...' pages 197–203

1 LMA ACC2558 NR 1/7
2 *Diary of Joseph Farington*, ed. Cave, K., (1983) vol.11, pp.3930–1, 15 May 1811
3 BAL, MyFam/9, Printed letter to Professor Robison, 1799
4 Farington, *Diary*, 4 March 1810; George Dance, *Portraits from the life*, 1814
5 *GM* vol. 81(i) 1811, p.499
6 TNA PROB 11/1523
7 BL Add. 36593 f. 172
8 George Dance, *Portraits from the life*, 1814. Earlier edition 1808
9 *Diary of Joseph Farington*, ed. Cave, K., (1983) vol.11, p.3930-1, 15 May 1811
10 Dennistoun, *Memoirs of Sir Robert Strange ... Lumisden*, James Bruce to Robert Strange, 16 May 1768
11 *Notes for the ordering of a 2nd trial, 67*, cited in Golan, op. cit. p.51.
12 LMA ACC2558 NR/1
13 James Elmes, *History of Architecture in Great Britain*, 1847
14 BCA MS 3219/4/39/6 11 February 1797, RM to James Watt
15 Nichols, John, *Literary anecdotes of the eighteenth century*, vol.9, 1815
16 Thacker, F. S., *The Thames Highway, a history of the inland navigation*, 1914, p.144
17 BAL MyFam 5/35, 5/36; BL Add 36593, f.172
18 Christopher Hussey, *English Country Houses, Mid Georgian 1760–1800*, 1956; *Country Life*, 1 February 1941
19 Colvin, *Biographical dictionary of British architects*
20 Survey of London, Ida Darlington, vol.xxv, *St George's Fields*, 1955, p.45
21 *Journal of Transport History*, first series, vol.2, 1955-56, No.3, May 1956, pp.191–2
22 BAL Diaries passim, MyFam/8
23 see Ruddock, Lindsay
23a Maudlin, Daniel, 'Robert Mylne at Pitlour House', *Architectural Heritage*, 2001, vol.xi, pp.27–37
24 Mylne, R.S., *The Master Masons to the Crown of Scotland*, 1893
25 Christie's, London, *Important architectural drawings and watercolours including the Sir Albert Richardson Collection*, 30 November 1983; Lots 17–60. Many went to the Canadian Centre for Architecture and the Getty Museum; see also in this context Richardson pp.39–45

Index

Note: Illustrations in the text are shown by page numbers in *italics*, those in the picture section have the prefix **Ill**.